Recent Trends in Pharmaceutical
Analytical Chemistry

Recent Trends in Pharmaceutical Analytical Chemistry

Editors

Constantinos K. Zacharis
Aikaterini Markopoulou

MDPI • Basel • Beijing • Wuhan • Barcelona • Belgrade • Manchester • Tokyo • Cluj • Tianjin

Editors
Constantinos K. Zacharis
Laboratory of Pharmaceutical
Analysis, Department of
Pharmaceutical Technology,
School of Pharmacy, Aristotle
University of Thessaloniki
Greece

Aikaterini Markopoulou
Laboratory of Pharmaceutical
Analysis, Department of
Pharmaceutical Technology,
School of Pharmacy, Aristotle
University of Thessaloniki
Greece

Editorial Office
MDPI
St. Alban-Anlage 66
4052 Basel, Switzerland

This is a reprint of articles from the Special Issue published online in the open access journal *Molecules* (ISSN 1420-3049) (available at: https://www.mdpi.com/journal/molecules/special_issues/pharmaceutical_analytical_chemistry).

For citation purposes, cite each article independently as indicated on the article page online and as indicated below:

LastName, A.A.; LastName, B.B.; LastName, C.C. Article Title. *Journal Name* **Year**, *Volume Number*, Page Range.

ISBN 978-3-0365-0798-9 (Hbk)
ISBN 978-3-0365-0799-6 (PDF)

© 2021 by the authors. Articles in this book are Open Access and distributed under the Creative Commons Attribution (CC BY) license, which allows users to download, copy and build upon published articles, as long as the author and publisher are properly credited, which ensures maximum dissemination and a wider impact of our publications.

The book as a whole is distributed by MDPI under the terms and conditions of the Creative Commons license CC BY-NC-ND.

Contents

About the Editors vii

Preface to "Recent Trends in Pharmaceutical Analytical Chemistry" ix

Constantinos K. Zacharis and Catherine K. Markopoulou
Recent Trends in Pharmaceutical Analytical Chemistry
Reprinted from: *Molecules* 2020, 25, 3560, doi:10.3390/molecules25163560 1

Wenrong Yao, Lei Yu, Wenhong Fan, Xinchang Shi, Lan Liu, Yonghong Li, Xi Qin, Chunming Rao and Junzhi Wang
A Cell-Based Strategy for Bioactivity Determination of Long-Acting Fc-Fusion Recombinant Human Growth Hormone
Reprinted from: *Molecules* 2019, 24, 1389, doi:10.3390/molecules24071389 5

Yang Song, Yu Zhou, Xiao-Ting Yan, Jing-Bo Bi, Xin Qiu, Yu Bian, Ke-Fei Wang, Yuan Zhang and Xue-Song Feng
Pharmacokinetics and Tissue Distribution of Alnustone in Rats after Intravenous Administration by Liquid Chromatography-Mass Spectrometry
Reprinted from: *Molecules* 2019, 24, 3183, doi:10.3390/molecules24173183 15

Yang Song, Yuan Zhang, Xiao-Yi Duan, Dong-Wei Cui, Xin Qiu, Yu Bian, Ke-Fei Wang and Xue-Song Feng
Pharmacokinetics and Tissue Distribution of Anwuligan in Rats after Intravenous and Intragastric Administration by Liquid Chromatography-Mass Spectrometry
Reprinted from: *Molecules* 2020, 25, 39, doi:10.3390/molecules25010039 31

Yang Song, Yuan Zhang, Wei-Peng Zhang, Bao-Zhen Zhang, Ke-Fei Wang and Xue-Song Feng
Interaction Effects between Doxorubicin and Hernandezine on the Pharmacokinetics by Liquid Chromatography Coupled with Mass Spectrometry
Reprinted from: *Molecules* 2019, 24, 3622, doi:10.3390/molecules24193622 45

Essam Ezzeldin, Muzaffar Iqbal, Yousif A. Asiri, Azza A Ali, Prawez Alam and Toqa El-Nahhas
A Hydrophilic Interaction Liquid Chromatography–Tandem Mass Spectrometry Quantitative Method for Determination of Baricitinib in Plasma, and Its Application in a Pharmacokinetic Study in Rats
Reprinted from: *Molecules* 2020, 25, 1600, doi:10.3390/molecules25071600 57

Nan Zhao, Hao-ran Tan, Qi-li Chen, Qi Sun, Lin Wang, Yang Song, Kamara Mohamed Olounfeh and Fan-hao Meng
Development and Validation of a Sensitive UHPLC-MS/MS Method for the Measurement of Gardneramine in Rat Plasma and Tissues and Its Application to Pharmacokinetics and Tissue Distribution Study
Reprinted from: *Molecules* 2019, 24, 3953, doi:10.3390/molecules24213953 71

Edvin Brusač, Mario-Livio Jeličić, Daniela Amidžić Klarić, Biljana Nigović, Nikša Turk, Ilija Klarić and Ana Mornar
Pharmacokinetic Profiling and Simultaneous Determination of Thiopurine Immunosuppressants and Folic Acid by Chromatographic Methods
Reprinted from: *Molecules* 2019, 24, 3469, doi:10.3390/molecules24193469 83

Paraskevas D. Tzanavaras, Sofia Papadimitriou and Constantinos K. Zacharis
Automated Stopped-Flow Fluorimetric Sensor for Biologically Active Adamantane Derivatives Based on Zone Fluidics
Reprinted from: *Molecules* **2019**, *24*, 3975, doi:10.3390/molecules24213975 . . . . **101**

Akane Funaki, Yuta Horikoshi, Teruyuki Kobayashi and Takashi Masadome
Determination of Polyhexamethylene Biguanide Hydrochloride Using a Lactone-Rhodamine B-Based Fluorescence Optode
Reprinted from: *Molecules* **2020**, *25*, 262, doi:10.3390/molecules25020262 . . . . **111**

Samir M. Ahmad, Mariana N. Oliveira, Nuno R. Neng and J.M.F. Nogueira
A Fast and Validated High Throughput Bar Adsorptive Microextraction (HT-BAμE) Method for the Determination of Ketamine and Norketamine in Urine Samples
Reprinted from: *Molecules* **2020**, *25*, 1438, doi:10.3390/molecules25061438 . . . . **119**

Eleni Tsanaktsidou, Christina Karavasili, Constantinos K. Zacharis, Dimitrios G. Fatouros and Catherine K. Markopoulou
Partial Least Square Model (PLS) as a Tool to Predict the Diffusion of Steroids Across Artificial Membranes
Reprinted from: *Molecules* **2020**, *25*, 1387, doi:10.3390/molecules25061387 . . . . **129**

Nao Wu, Stéphane Balayssac, Saïda Danoun, Myriam Malet-Martino and Véronique Gilard
Chemometric Analysis of Low-field ^{1}H NMR Spectra for Unveiling Adulteration of Slimming Dietary Supplements by Pharmaceutical Compounds
Reprinted from: *Molecules* **2020**, *25*, 1193, doi:10.3390/molecules25051193 . . . . **145**

About the Editors

Constantinos K. Zacharis currently an Assistant Professor in the School of Pharmacy (Aristotle University of Thessaloniki, Greece). Dr. Zacharis received his BSc in Chemistry in 2001 and his PhD degree in Analytical Chemistry in 2006 from the Department of Chemistry (AUTh). Since 2010, he has held the position of Editor of the Analytical Chemistry section of the international scientific journal Open Chemistry (formerly Central European Journal of Chemistry) and serves on the editorial boards of Molecules, SCI, Analytica, SEJ Pharmaceutical Analysis, Instrumentation Science & Technology, Advances in Analytical Chemistry and Current Analysis on Chemistry. He has authored and co-authored more than 60 scientific articles (H-index = 20/Scopus) and has reviewed more than 700 papers in 50 scientific Journals. He has also acted as Guest Editor in almost 10 Special Issues. His current research is mainly focused on the development and application of novel analytical methodologies using separation techniques (LC, GC, CE) for pharmaceutical analysis.

Aikaterini Markopoulou Since 2015, Catherine has been an Associate Professor of Pharmaceutical Analysis in the School of Pharmacy (Aristotle University of Thessaloniki, Greece). She has authored more than 53 research articles (H-index = 14, Scopus) and supervised 3 PhD and 11 post-graduate students. Her scientific interests involve: (i) method development for the assay of active substances in pharmaceutical formulations, (ii) study of the stability for active substances in various pharmaceutical formulation, (iii) analysis of drugs in biological fluids and various substrates, by LC/MS-ESI, (iv) modeling and development of multi-parameter systems for studying the retention mechanism of analytes in reverse phase HPLC columns and (v) the study of drug-likeness by modeling and HPLC. As a research team member, Catherine has participated in five funded research projects

Preface to "Recent Trends in Pharmaceutical Analytical Chemistry"

We are proud to announce the launch of the present Special Issue on "Recent Trends in Pharmaceutical Analytical Chemistry" published in the international journal, Molecules.Modern analytical chemistry plays a vital role in pharmaceutical science. It provides precise and accurate data supporting processes related to drug discovery and development. Some paradigms comprise the purity of drug substances during synthesis, pharmacokinetic studies, drug stability, elucidation of the drug metabolic pathways, drug–protein interactions, etc. New methodologies, state-of-the art instrumentation and materials, and automated systems—offering precise and accurate analytical data—have become essential prerequisites in this field. Moreover, chemometrics have proven to be a beneficial tool for the optimization of method parameters and can also be used to identify and minimize the sources of variability that may lead to poor method robustness.The participation of colleagues from all over the world has been an impressive feat, and following the peer-review and evaluation phases of the articles, this Special Issue is finally composed of twelve up-to-date research articles covering various topics. We hope readers will find the publications interesting and informative.As Guest Editors of this Special Issue, we are sincerely grateful to all the contributors for their outstanding cooperation and valuable support. Special thanks are also warranted to all the peer reviewers for their valuable suggestions and criticisms that were instrumental in improving the quality of manuscripts.

Constantinos K. Zacharis, Aikaterini Markopoulou
Editors

Editorial

Recent Trends in Pharmaceutical Analytical Chemistry

Constantinos K. Zacharis * and Catherine K. Markopoulou

Laboratory of Pharmaceutical Analysis, Department of Pharmaceutical Technology, School of Pharmacy, Aristotle University of Thessaloniki, 54124 Thessaloniki, Greece; amarkopo@pharm.auth.gr
* Correspondence: czacharis@pharm.auth.gr

Received: 6 July 2020; Accepted: 10 July 2020; Published: 5 August 2020

Modern analytical chemistry plays a vital role in pharmaceutical sciences. It provides precise and accurate data supporting the processes related to drug discovery and development. Some examples include the purity of drug substances during synthesis, pharmacokinetic studies, drug stability, elucidation of the drug metabolic pathways, drug–protein interactions, etc. New methodologies, state-of-the art instrumentation and materials, and automated systems—offering precise and accurate analytical data—have become essential prerequisites in this field. Moreover, chemometrics have been proved to be a beneficial tool for the optimization of method parameters and can also identify and minimize the sources of variability that may lead to poor method robustness.

The present Special Issue comprises twelve full-length research articles covering the latest research trends and applications of pharmaceutical analytical chemistry. An interesting cell-based bioassay was proposed by the research group of C. Rao and J. Wang for the determination of the bioactivity of long-acting growth hormone as a potential alternative biopharmaceutical for the treatment of growth hormone deficiency [1]. The method is based on the usage of a luciferase reporter gene system, which is involved in the full-length human GH receptor (hGHR) and the SIE and GAS (SG) response element. The proposed protocol can be used in routine analysis and it was validated according to ICH guidelines and the Chinese Pharmacopoeia.

The research group of X.-S. Feng presented *"Pharmacokinetics and Tissue Distribution of Alnustone in Rats after Intravenous Administration by Liquid Chromatography-Mass Spectrometry"* [2]. A simple and fast LC-MS/MS method was developed to evaluate the pharmacokinetic and tissue distribution profiles of alnustone in rats. The sample preparation involved simple protein precipitation using acetonitrile. After the optimization of LC-MS/MS conditions, the method was fully validated according to the bioanalytical method validation guidance for industry (FDA). The authors concluded that alnustone is predominantly distributed in the lung and liver tissues within 1 h and therefore these tissues might be the target organs for the curative effect of the drug. Analogous research has been published by the same authors related to the *"Pharmacokinetics and Tissue Distribution of Anwuligan in Rats after Intravenous and Intragastric Administration by Liquid Chromatography-Mass Spectrometry"* [3]. In contrast to their previous work [2], a liquid–liquid extraction-based sample preparation protocol was developed and optimized for the isolation of anwuligan from rat plasma and tissues with minimum matrix effects. According to their findings, the half-time elimination of the drug was relatively shorter after intragastric administration compared to the intravenous one. An interesting investigation was also published by the same researchers in their publication *"Interaction Effects between Doxorubicin and Hernandezine on the Pharmacokinetics by Liquid Chromatography Coupled with Mass Spectrometry"* [4]. One of the primary targets of the work was the development of an LC-MS/MS method for the determination of both drugs in rat plasma and also to compare their pharmacokinetic data after intravenous administration in rats. According to their findings, there were significant differences in the pharmacokinetic profiles—especially in the C_{max} and $AUC_{0-\infty}$ parameters of doxorubicin—indicating that hernandezine could improve the absorption of the anticancer drug. The outcome of this research might be used as clinical guidance for doxorubicin and hernandezine to prevent adverse reactions.

Dr. El-Nahhas et al., in their interesting contribution *"A Hydrophilic Interaction Liquid Chromatography–Tandem Mass Spectrometry Quantitative Method for Determination of Baricitinib in Plasma, and Its Application in a Pharmacokinetic Study in Rats"*, presented quantitation data from the determination of baricitinib in rat plasma [5]. A liquid–liquid extraction procedure with two extraction solvents (n-hexane and dichloromethane) was proposed for drug isolation from biological matrices, avoiding the potential enhancement or suppression of the MS signal. The authors investigated the separation of baricitinib and the internal standard under HILIC conditions, achieving a 3-min isocratic separation.

A sensitive UHPLC-MS/MS method has been published by F.-H. Meng et al. in their interesting article entitled *"Development and Validation of a Sensitive UHPLC-MS/MS Method for the Measurement of Gardneramine in Rat Plasma and Tissues and Its Application to Pharmacokinetics and Tissue Distribution Study"* [6]. Gardneramine is a monoterpenoid indole alkaloid with an exceptional central depressive effect and actions on myocardium. The pharmacokinetic properties and tissue distribution of the selected drug have been thoroughly investigated after its intravenous administration. To support this task, the authors developed a fast UHPLC-MS/MS method in combination with simple protein precipitation for sample preparation.

Pharmacokinetic profiling data of thiopurine immunosuppressants and folic acid were provided by the research group of Mornar in the report entitled *"Pharmacokinetic Profiling and Simultaneous Determination of Thiopurine Immunosuppressants and Folic Acid by Chromatographic Methods"* [7]. The authors evaluated the pharmacokinetic profiling of azathioprine, 6-mercaptopurine, 6-thioguanine, and folic acid using various chromatographic techniques and in silico methodology. An HPLC method was also developed for the simultaneous determination of the analytes in their commercially available formulations.

An automated stopped-flow fluorimetric sensor based on zone fluidics was proposed by P. D. Tzanavaras et al. in their work *"Automated Stopped-Flow Fluorimetric Sensor for Biologically Active Adamantane Derivatives Based on Zone Fluidics"* [8]. One of the main scopes of the research was to develop an automated method for the fast determination of amantadine, memantine, and rimantadine for the quality control of their commercially available formulations. Interestingly, an amantadine-containing formulation obtained from a third—non-EU—country was found to be out of specification. These results were confirmed using a validated HPLC method.

Dr. Masadome and co-workers, in their article *"Determination of Polyhexamethylene Biguanide Hydrochloride Using a Lactone-Rhodamine B-Based Fluorescence Optode"*, reported a new approach for the quantitation of polyhexamethylene biguanide using a lactone–rhodamine B fluorescence optode [9]. The analyte is a cationic polyelectrolyte which is used for disinfectants in contact lens detergents and sanitizers for swimming pools. The principle of the method is based on the fluorescence quenching of a lactone–rhodamine B-containing optode membrane caused by the analyte. The method is quite simple without using toxic reagents.

A high-throughput bar adsorptive microextraction for the analysis of ketamine and norketamine in urine has been proposed by J.M.F. Nogueira et al. in their interesting publication *"A Fast and Validated High Throughput Bar Adsorptive Microextraction (HT-BAμE) Method for the Determination of Ketamine and Norketamine in Urine Samples"* [10]. The identification and quantification of both drugs were performed using large-volume injection gas chromatography–mass spectrometry operating in the selected ion monitoring mode. A set of parameters that could influence the performance of the microextraction was investigated and optimized. The proposed analytical scheme has the possibility of performing parallel microextractions and the subsequent desorption of up to 100 samples in a single apparatus in just 45 min.

Dr. Markopoulou et al., in their study *"Partial Least Square Model (PLS) as a Tool to Predict the Diffusion of Steroids across Artificial Membranes"*, presented data regarding the ability of a drug to permeate human tissues using a partial least square-based statistical approach [11]. The researchers attempted to decode the drug permeability by correlating the apparent permeability coefficient of 33 steroids with their physicochemical and structural properties. The apparent permeability coefficient of the compounds was determined by in vitro experiments. The obtained models could be utilized to predict the permeability of a new drug candidate without animal experiments.

Last, but not least, Drs. Balayssac and Gilard and their colleagues in the publication *"Chemometric Analysis of Low-field 1H NMR Spectra for Unveiling Adulteration of Slimming Dietary Supplements by Pharmaceutical Compounds"* presented the development of a novel chemometric-based NMR approach for unveiling the adulteration of slimming dietary supplements [12]. The authors analyzed 66 dietary supplements using low-field ^{1}H-NMR to build the PLS model. The potential limitations of the proposed method consist of the poor sensitivity and the low spectral resolution of the low-field ^{1}H-NMR instrument.

As Guest Editors of this Special Issue, we are sincerely grateful to all the contributors for their outstanding cooperation and valuable support. Special thanks are also due to all the peer reviewers for their valuable suggestions and criticisms.

Funding: This research received no external funding.

Conflicts of Interest: The authors declare no conflict of interest.

References

1. Yao, W.; Yu, L.; Fan, W.; Shi, X.; Liu, L.; Li, Y.; Qin, X.; Rao, C.; Wang, J. A Cell-Based Strategy for Bioactivity Determination of Long-Acting Fc-Fusion Recombinant Human Growth Hormone. *Molecules* **2019**, *24*, 1389. [CrossRef] [PubMed]
2. Song, Y.; Zhou, Y.; Yan, X.-T.; Bi, J.-B.; Qiu, X.; Bian, Y.; Wang, K.-F.; Zhang, Y.; Feng, X.-S. Pharmacokinetics and Tissue Distribution of Alnustone in Rats after Intravenous Administration by Liquid Chromatography-Mass Spectrometry. *Molecules* **2019**, *24*, 3183. [CrossRef] [PubMed]
3. Song, Y.; Zhang, Y.; Duan, X.-Y.; Cui, D.-W.; Qiu, X.; Bian, Y.; Wang, K.-F.; Feng, X.-S. Pharmacokinetics and Tissue Distribution of Anwuligan in Rats after Intravenous and Intragastric Administration by Liquid Chromatography-Mass Spectrometry. *Molecules* **2019**, *25*, 39. [CrossRef] [PubMed]
4. Song, Y.; Zhang, Y.; Zhang, W.-P.; Zhang, B.-Z.; Wang, K.-F.; Feng, X.-S. Interaction Effects between Doxorubicin and Hernandezine on the Pharmacokinetics by Liquid Chromatography Coupled with Mass Spectrometry. *Molecules* **2019**, *24*, 3622. [CrossRef] [PubMed]
5. Ezzeldin, E.; Iqbal, M.; Asiri, Y.A.; Ali, A.A.; Alam, P.; El-Nahhas, T. A Hydrophilic Interaction Liquid Chromatography–Tandem Mass Spectrometry Quantitative Method for Determination of Baricitinib in Plasma, and Its Application in a Pharmacokinetic Study in Rats. *Molecules* **2020**, *25*, 1600. [CrossRef] [PubMed]
6. Zhao, N.; Tan, H.R.; Chen, Q.L.; Sun, Q.; Wang, L.; Song, Y.; Olounfeh, K.M.; Meng, F.H. Development and Validation of a Sensitive UHPLC-MS/MS Method for the Measurement of Gardneramine in Rat Plasma and Tissues and its Application to Pharmacokinetics and Tissue Distribution Study. *Molecules* **2019**, *24*, 3953. [CrossRef] [PubMed]
7. Brusač, E.; Jeličić, M.L.; Amidžić Klarić, D.; Nigović, B.; Turk, N.; Klarić, I.; Mornar, A. Pharmacokinetic Profiling and Simultaneous Determination of Thiopurine Immunosuppressants and Folic Acid by Chromatographic Methods. *Molecules* **2019**, *24*, 3469. [CrossRef] [PubMed]
8. Tzanavaras, P.D.; Papadimitriou, S.; Zacharis, C.K. Automated Stopped-Flow Fluorimetric Sensor for Biologically Active Adamantane Derivatives Based on Zone Fluidics. *Molecules* **2019**, *24*, 3975. [CrossRef] [PubMed]
9. Funaki, A.; Horikoshi, Y.; Kobayashi, T.; Masadome, T. Determination of Polyhexamethylene Biguanide Hydrochloride Using a Lactone-Rhodamine B-Based Fluorescence Optode. *Molecules* **2020**, *25*, 262. [CrossRef] [PubMed]
10. Ahmad, S.M.; Oliveira, M.N.; Neng, N.R.; Nogueira, J.M.F. A Fast and Validated High Throughput Bar Adsorptive Microextraction (HT-BAμE) Method for the Determination of Ketamine and Norketamine in Urine Samples. *Molecules* **2020**, *25*, 1438. [CrossRef] [PubMed]

11. Tsanaktsidou, E.; Karavasili, C.; Zacharis, C.K.; Fatouros, D.G.; Markopoulou, C.K. Partial least square model (PLS) as a tool to predict the diffusion of steroids across artificial membranes. *Molecules* **2020**. [CrossRef] [PubMed]
12. Wu, N.; Balayssac, S.; Danoun, S.; Malet-Martino, M.; Gilard, V. Chemometric Analysis of Low-field 1H NMR Spectra for Unveiling Adulteration of Slimming Dietary Supplements by Pharmaceutical Compounds. *Molecules* **2020**, *25*, 1193. [CrossRef] [PubMed]

© 2020 by the authors. Licensee MDPI, Basel, Switzerland. This article is an open access article distributed under the terms and conditions of the Creative Commons Attribution (CC BY) license (http://creativecommons.org/licenses/by/4.0/).

Article

A Cell-Based Strategy for Bioactivity Determination of Long-Acting Fc-Fusion Recombinant Human Growth Hormone

Wenrong Yao [†], Lei Yu [†], Wenhong Fan [†], Xinchang Shi, Lan Liu, Yonghong Li, Xi Qin, Chunming Rao * and Junzhi Wang *

National Institutes for Food and Drug Control, No. 2, Tiantan Xili, Beijing 100050, China; yz1322@126.com (W.Y.); yulei@nifdc.org.cn (L.Y.); fanwh@nifdc.org.cn (W.F.); shixc@nifdc.org.cn (X.S.); liulan@nifdc.org.cn (L.L.); liyh@nifdc.org.cn (Y.L.); qinxi@nifdc.org.cn (X.Q.)
* Correspondence: raocm@nifdc.org.cn (C.R.); wangjz_nifdc2014@163.com (J.W.); Tel.: +86-10-67095380 (C.R.); +86-10-67095782 (J.W.)
† These authors contributed equally to this work.

Received: 27 February 2019; Accepted: 8 April 2019; Published: 9 April 2019

Abstract: The long-acting growth hormone (LAGH) is a promising alternative biopharmaceutical to treat growth hormone (GH) deficiency in children, and it was developed using a variety of technologies by several pharmaceutical companies. Most LAGH preparations, such as Fc fusion protein, are currently undergoing preclinical study and clinical trials. Accurate determination of bioactivity is critical for the efficacy of quality control systems of LAGH. The current in vivo rat weight gain assays used to determine the bioactivity of recombinant human GH (rhGH) in pharmacopoeias are time-consuming, expensive, and imprecise, and there are no recommended bioassays for LAGH bioactivity in pharmacopoeias. Therefore, we developed a cell-based bioassay for bioactivity determination of therapeutic long-acting Fc-fusion recombinant human growth hormone (rhGH-Fc) based on the luciferase reporter gene system, which is involved in the full-length human GH receptor (hGHR) and the SG (SIE and GAS) response element. The established bioassay was comprehensively validated according to the International Council for Harmonization (ICH) Q2 (R1) guidelines and the Chinese Pharmacopoeia, and is highly precise, time-saving, simple, and robust. The validated bioassay could be qualified for bioactivity determination during the research, development, and manufacture of rhGH-Fc, and other LAGH formulations.

Keywords: growth hormone; long-acting Fc-fusion recombinant human growth hormone; method validation; cell-based bioassay; reporter gene assay

1. Introduction

Growth hormone (GH), also known as somatotropin, is pituitary-derived anabolic hormone that regulates a number of metabolic processes involved in growth and development [1,2]. The specific responses to body growth and metabolism act via multiple GH signal transduction pathways. Upon binding to the GH receptor, the GH-associated janus kinase 2 (JAK2) tyrosine kinase is activated, and then the signaling proteins to GH receptor (GHR)–JAK2 complexes are recruited by phosphorylated tyrosines and subsequently induce signal transducers and activators of transcription (STAT), mitogen-activated protein kinases (MAPK), and phosphatidylinositol 3 kinase (PI3K) intracellular signaling pathways [3,4]. GH deficiency (GHD) causes growth retardation in children and metabolic dysfunction in adults. Recombinant human GH (rhGH) is a 22 kDa polypeptide that consists of 191 amino acids. Treatment with rhGH in children with GHD has been well established since its approval by the United States Food and Drug Administration (FDA) in 1985 [1,5].

rhGH has a short-circulating half-life of only a few hours in the human body, and the therapeutic regimens rely on daily subcutaneous injections that can be burdensome, inconvenient, and painful for some patients. Further, daily injections may cause non-adherence over time and result in decreased therapeutic efficacy [6–8]. Therefore, long-acting growth hormone (LAGH) formulations with extended half-lives have been developed to improve adherence and simplify dosing schedules, which were existing disadvantages of rhGH [6,8–13]. Regarding the LAGH formulations, several forms have been developed by increasing the effective size of rhGH and reducing its rate of clearance from the body, such as Fc fusion rhGH and PEGylated rhGH. Additionally, Fc domain binding to the neonatal Fc receptor (FcRn) prolongs the serum half-life of rhGH-Fc, and salvages the protein from being degraded in endosomes [7,14–16].

Due to its unique pharmacokinetics and pharmacodynamics, and inherent complexity, the functional characterization of LAGH is especially required to closely monitor bioactivity, bioidentity, and other types of physicochemical characterization [6,9,11,16,17]. Nowadays, the bioassay for rhGH bioactivity highly relies on in vivo rat weight gain, as described in the US Pharmacopeia and the Chinese Pharmacopoeia, and it is time-consuming, expensive, and imprecise. Additionally, the in vitro rat Nb2-11 cell-based proliferation assay is recorded in the US Pharmacopeia. The limitation of the proliferation assay is that the lactogenic bioactivity of rhGH, rather than the bioactivity of rhGH, is determined through responsiveness to prolactin receptors expressed on the Nb2-11 surface [18]. In regard to the bioactivity of LAGH, there are no recommended bioassays in the Chinese Pharmacopoeia, although the PEGylated-rhGH (Jintrolong) LAGH was commercially developed and marketed in China [7]. Consequently, the development of an animal-free bioassay with a low coefficient of variation (CV) that is time-saving and simple to carry out needs to be explored to facilitate the determination of bioactivity for LAGH formulations. The unchanged specificity, lower variability, and simpler procedures of reporter gene assays (RGAs) compared with animal assays render them excellent alternatives for the bioactivity determination of biopharmaceuticals [19,20]. Actually, RGAs focused on the hGHR-LHRE reporter system are suitable for the measurement of the biological activity of serum GH, rhGH, and ligand–receptor fusion of GH. However, transient transfection and the lack of method validation have probably impeded the drive to determine the bioactivity of therapeutic rhGH or LAGH [18,21–23].

Here, we develop a cell-based bioassay to determine the bioactivity of therapeutic rhGH-Fc using RGA and comprehensively validate it according to the International Council for Harmonization (ICH) Q2 (R1) guidelines and the Chinese Pharmacopoeia.

2. Materials and Methods

2.1. Cells and Materials

The human embryonic kidney 293 (HEK293) cell line (CRL-1573™) was obtained from the American Type Culture Collection (Manassas, VA, USA). RPMI1640, fetal bovine serum (FBS), Geneticin (G418), and hygromycin B were purchased from Gibco (Grand Island, NY, USA). Human growth hormone receptor (hGHR) plasmid was obtained from OriGene Technologies (Beijing, China). ViaFect™ transfection reagent was purchased from Promega (Madison, WI, USA). The Britelite Plus Reporter Gene Assay System was obtained from PerkinElmer (Waltham, MA, USA). The SG-luciferase reporter plasmid (including SIE and GAS response elements) [19], an in-house reference of rhGH-Fc, rhGH-Fc, rhGH, Fc-fusion recombinant human erythropoietin (rhEPO-Fc), Fc-fusion vascular endothelial growth factor receptor (VEGFR-Fc), Fc-fusion recombinant human interleukin 15 (rhIL15-Fc), and Fc fusion recombinant human GLP1 (rhGLP-Fc) were stored at 4 °C or −80 °C in our laboratory. The 96-well flat clear bottom white polystyrene TC-treated microplates were purchased from Corning (New York, NY, USA).

2.2. Preparation of Responsive Cells to rhGH-Fc

The HEK293 cells were cultured in DMEM supplemented with 10% FBS in an incubator atmosphere of 5% CO_2 at 37 °C, and they were split every 3 days. The SG-luciferase reporter plasmid and hGHR plasmid were transfected simultaneously into HEK293 cells using ViaFect™ transfection reagent according to the manufacturer's instructions. The cells expressing SG-luciferase and hGHR received regular changes of DMEM-10% FBS with hygromycin B (300 μg/mL) and G418 (1000 μg/mL) and were continuously cultured for at least for 5 weeks. Then, a clonal cell line derived from a single cell was produced by limiting the dilution from the stably transfected HEK293 cells using 0.5 cells per well in a 96-well plate. After isolating the clones, assessments through rhGH-Fc stimulation were performed. The cells, namely HEK293-Luc, were highly responsive to rhGH-Fc. The HEK293-Luc cells were maintained in DMEM-10% FBS with 200 μg/mL of hygromycin B and 500 μg/mL of G418. Meanwhile, the HEK293 cell line only transfected with SG-luciferase plasmid without hGHR plasmid was used as the control group.

2.3. Bioactivity Assay

The bioassay of rhGH-Fc was performed in vitro as a cell-based bioassay as described previously [20]. In brief, 4×10^4 cells in 70 μL RPMI1640 were added to each well in a 96-well cell plate, and they were incubated for 18–22 h at 37 °C in a 5% CO_2 incubator. An rhGH-Fc (30 mg/mL) and in-house rhGH-Fc reference (31.4 mg/mL) were diluted by serial 3-fold dilutions with RPMI1640 from pre-diluted starting concentrations of 30,000 ng/mL. Then, 70 μL of diluted rhGH-Fc was added to cells. It should be noted that the final working concentration of rhGH-Fc was 15,000 ng/mL, and the final concentration ranged from 6.9 ng/ml to 15,000 ng/mL. After 5 h of incubation at 37 °C in a 5% CO_2 incubator, the supernatants were removed, and 60 μL Britelite Plus Reporter Gene Assay reagent was added to each well. After 5 min of mixing the reagents under subdued light conditions, the relative luciferase units (RLU) were determined by a Luminoscan Ascent plate reader (Molecular Devices, San Jose, CA, USA).

The sigmoidal curves, signal-noise-ratio (SNR), and 50% effective concentration (EC_{50}) were calculated through a four-parameter method. The SNR was indicated by the ratio of the RLU of the top asymptote to the bottom asymptote. The relative bioactivity of rhGH-Fc was shown as the ratio of the EC_{50} values of an in-house reference to the EC_{50} values of the rhGH-Fc samples.

2.4. Preparation of Forced Degradation of rhGH-Fc

The forced degradation of rhGH-Fc was applied under thermal stress conditions as described previously [20]. The rhGH-Fc samples were kept at 60 °C for 1, 3, 5, and 7 days to induce degradation. After that, the degraded rhGH-Fc was used to determine the bioactivity according to the procedure given under Section 2.3.

2.5. Data Analysis and Statistics

All the statistical tests were performed using GraphPad Prism 7.0 (GraphPad Software Inc., San Diego, CA, USA). Comparisons between two groups were performed using a two-tailed Mann–Whitney test, and multiple comparisons were performed using a Kruskal–Wallis test with Dunn's multiple comparisons. p-values < 0.05 were deemed to be statistically significant. Each experiment was performed in triplicate. The RLU of each concentration indicates the mean of three replicates. The mean ± SD is shown on each curve.

3. Results

3.1. Generation of Stable rhGH-Fc-Responsive Cell Line

To obtain a stable rhGH-Fc-responsive cell line for the determination of the bioactivity of rhGH-Fc, transfected HEK293 cells expressing SG-luciferase and hGHR (SG-hGHR), and SG-luciferase without hGHR (SG) were constructed. The two cell lines were evaluated for luciferase activity after rhGH-Fc stimulation. The pre-diluted starting concentration of rhGH-Fc was 50,000 ng/mL, and then it was diluted with a dilution ratio of 1:5. Figure 1A showed that sigmoidal curves increased dose-dependently with increasing rhGH-Fc, and the cells expressing SG-hGHR had a better SNR (SNR of 104.0) than control cells that did not express hGHR (SNR of 2.2). In the fourteen clonal cell lines from single cells obtained by limiting dilutions (Figure 1B), a representative cell line, 1B1 (named HEK293-Luc), was chosen for further method validation due to having the highest SNR (SNR of 221.9), a low EC_{50} (207.9 ng/mL), and rhGH-Fc-dependent behavior.

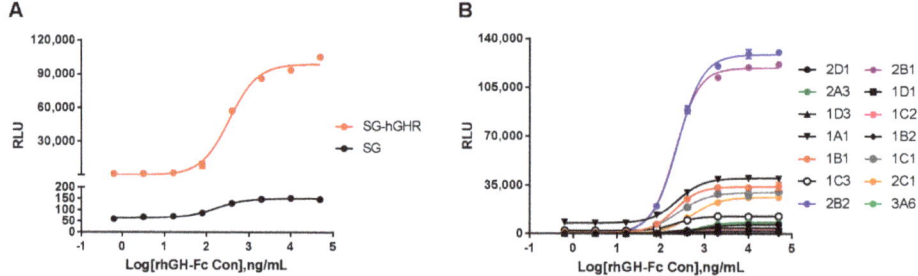

Figure 1. The responsiveness of transfected HEK293 cells to Fc-fusion recombinant human growth hormone (rhGH-Fc). The signal-to-noise ratio (SNR) of HEK293 cells expressing SG-hGHR was significantly higher than that of the SG cells (A). Fourteen clonal cell lines expressing SG-hGHR were tested (B). The establishment of a responsive cell line to rhGH-Fc was evaluated based on its 50% effective concentration (EC_{50}) and SNR. SG-hGHR, SIE and GAS response elements (SG) and human growth hormone receptor (hGHR).

3.2. Method Development and Optimization

To optimize the bioassay conditions, three parameters were investigated: the initial concentration of rhGH-Fc, the number of cells per well, and the incubation time. First, the linear region of the dose-response curve of rhGH-Fc was determined by luciferase activity induced by rhGH-Fc and was set at approximately 30.5–15,625 ng/mL (data not shown). The pre-diluted starting concentration of rhGH-Fc was set from 10,000 ng/mL to 25,000 ng/mL with three dilution factors. All of the sigmoidal curves illustrated a dose-dependent trend, and the EC_{50} (276.0–309.5 ng/mL) was similar for all four pre-diluted starting concentrations (Figure 3A). According to the SNR and uniformity of the concentration points, the optimal pre-diluted starting concentration was 15,000 ng/mL (SNR of 19.6). As can be seen from Figure 2B, all of the dose-dependent curves and EC_{50} values were nearly identical (334.4–389.7 ng/mL). The highest magnitude of RLU was illustrated between 4×10^4 and 5×10^4 cells per well, and a higher SNR (SNR of 330.6) was observed with 4×10^4 cells per well. We therefore chose 4×10^4 cells per well as the optimal cell number. Finally, No RLU of the top asymptote increased with an increasing incubation time, and no significant differences were observed in the EC_{50} between 4 and 6 h. Due to its higher SNR (SNR of 107.5), 5 h was deemed to be an applicable incubation time under the given conditions (Figure 2C).

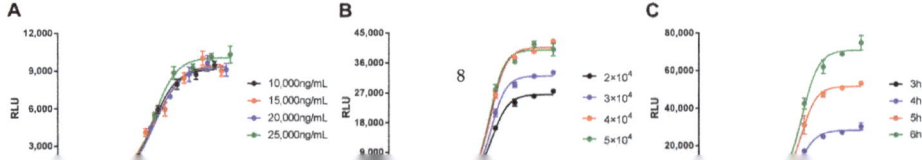

values were nearly identical (334.4–380.7 ng/mL). The highest magnitude of RLU was illustrated between 4×10^4 and 5×10^4 cells per well, and a higher SNR (SNR of 210.6) was observed with 4×10^4 cells per well. We therefore chose 4×10^4 cells per well as the optimal cell number. Finally, the RLU of the top asymptote increased with an increasing incubation time, and no significant differences were observed in the EC_{50} between 4 and 6 h. Due to its higher SNR (SNR of 107.5), 5 h was deemed to be an applicable incubation time under the given conditions (Figure 2C).

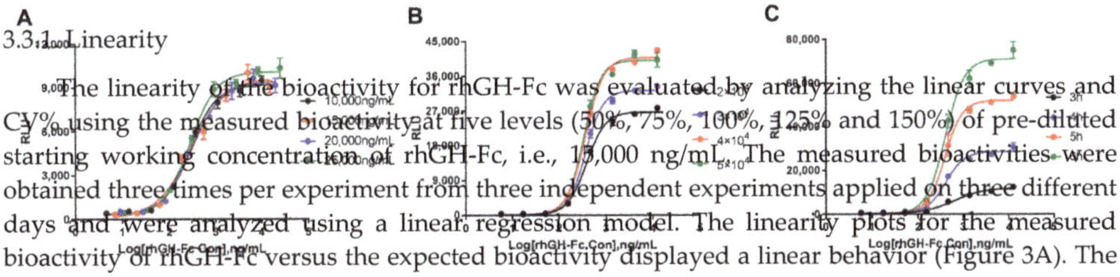

Figure 2. Optimization of experimental parameters of the bioassay. (**A**) Final working concentration of rhGH-Fc. (**B**) Cell numbers per well. (**C**). Incubation time with HEK293-Luc cells and rhGH-Fc.

3.3. *Method Validation*

A comprehensive validation of the bioassay was performed according to the ICH Q2 (R1) guidelines and the Chinese Pharmacopoeia [24,25].

3.3.1. Linearity

The linearity of the bioactivity for rhGH-Fc was evaluated by analyzing the linear curves and CV% using the measured bioactivity at five levels (50%, 75%, 100%, 125% and 150%) of pre-diluted starting working concentration of rhGH-Fc, i.e., 15,000 ng/mL. The measured bioactivities were obtained three times per experiment from three independent experiments applied on three different days and were analyzed using a linear regression model. The linearity plots for the measured bioactivity of rhGH-Fc versus the expected bioactivity displayed a linear behavior (Figure 3A). The CV% of measured bioactivity at five level concentrations ranged from 1.32% to 6.47%, i.e., less than 7.00%.

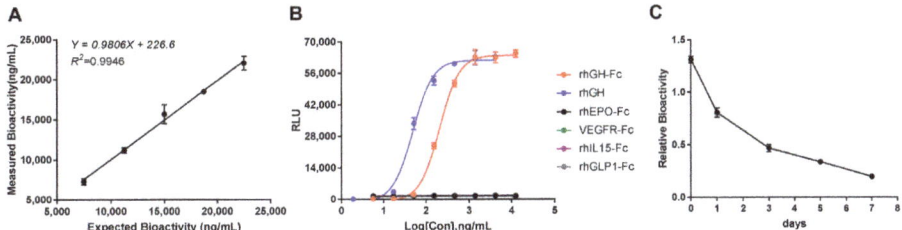

Figure 3. The validation of linearity and specificity. (**A**) A linearity plot of the expected bioactivity against the measured bioactivity. (**B**) The specificity of the bioassay to rhGH-Fc, rhGH and other therapeutic Fc fusion proteins. (**C**) The specificity of the bioassay to the thermal degradation of rhGH-Fc. The representative linearity plots of three independent experiments are shown.

3.3.2. Precision

The intra-day and inter-day precision were estimated by analyzing the relative bioactivity of rhGH-Fc samples in triplicate on the same day (intra-day precision) for three different days (inter-day precision). A batch of the final rhGH-Fc product and three freeze-thawed final rhGH-Fc samples were used for this purpose. The precision was expressed by CV% (shown in Table 1). The CV% of intra-day and inter-day precision were lower than 9.00% and 6.00%, respectively, indicating good precision and repeatability of the bioassay. The intermediate precision was determined by a comparison of relative bioactivity performed by two analysts on different days. The relative bioactivity of the seven final rhGH-Fc samples provided a consistent performance, as assessed by the two analysts (the mean relative bioactivity was 1.06 for analyst A, and 1.09 for analyst B). Moreover, the intra-plate of precision based on five repeated analyses for one rhGH-Fc sample in the same 96-well plate was 4.49%.

Table 1. The precision of the rhGH-Fc bioactivity results.

	Intra-Day CV%			Inter-Day CV%	95% CI [1] of Mean Relative Bioactivity
	1	2	3		
rhGH-Fc	2.42	3.82	5.00	2.06	1.10–1.17
rhGH-Fc D1 [2]	2.71	4.16	5.96	4.74	1.00–1.09
rhGH-Fc D3 [2]	2.04	5.05	8.81	2.27	0.80–0.87
rhGH-Fc D5 [2]	7.60	3.01	2.16	5.87	0.93–1.03

[1] CI: confidence interval. [2] The final rhGH-Fc product was frozen at −80 °C and thawed at 25 °C for 1, 3, and 5 cycles, respectively.

3.3.3. Accuracy

The accuracy of the described method was evaluated by the definite addition of an in-house reference in a sample, and it is indicated in terms of recovery (%) [20]. Here, the recovery was calculated using the rhGH-Fc spiking method at 50% of the level of the in-house reference concentration. The assay was performed three times per day on three different days. The mean recovery of the in-house reference was 113.64%, and the 95% confidence interval (CI) of that was 106.00–121.30%. The maximum CV% of the intra-day precision was 7.79%, and that of the inter-day precision was 8.73% (Table 2).

Table 2. The results of recovery for an in-house rhGH-Fc reference.

	Recovery (%)			Intra-day CV%	Mean	SD *
	1	2	3			
1	127.21	130.54	112.49	7.79	123.41	9.61
2	112.69	115.19	101.98	6.38	109.95	7.02
3	100.96	111.20	110.52	5.32	107.56	5.73
Mean		113.64				
SD *		9.92				
Inter-day CV%		8.73				

* SD: Standard deviation.

3.3.4. Specificity

The specificity of the method was confirmed by the responsiveness of HEK293-Luc cells to other Fc fusion therapeutic drugs, e.g., rhEPO-Fc, VEGFR-Fc, rhIL15-Fc, and rhGLP1-Fc (Figure 3B). None of them had any effect on HEK293-Luc cells. On the other hand, an expected result for rhGH with a standard sigmoidal curve parallel to rhGH-Fc was observed. Further, we forced the degradation of rhGH-Fc by thermal stress [20,24,26]. After incubation at 60 °C for different number of days, we saw a gradual decline in the relative bioactivity of rhGH-Fc as the number of days increased, which ranged from 1.31 in the normal sample to 0.19 in the degraded rhGH-Fc sample at 7 days (Figure 3C).

3.3.5. Robustness

To investigate the robustness of the bioassay, some changes to the assay media were made. As shown in Figure 4A, the responsiveness of HEK293-Luc cells to rhGH-Fc in FBS-free RPMI 1640 was observed. The RLU values of top asymptotes in assay media supplemented with 10%, 0.5%, and 0% FBS decreased gradually. No significant changes in the SNR were found because of the different RLU values of the bottom asymptotes (the mean RLUs were 1425.4, 896.8, and 675.7, respectively). Additionally, the results of the stability of HEK293-Luc cells are shown in Figure 4B,C. Parallel dose–response sigmoidal curves were found between passage 5 and passage 45 (Figure 4B). The EC_{50} values of passage 5, passage 16, and passage 45 were similar: 141.1, 145.3, and 147.9 ng/mL, respectively. The SNR of passage 16 was significantly higher than that of passage 5 ($p = 0.0087$), having changed from 57.13 to 79.90. As expected, the SNR of passage 45 (72.48) was consistent with that of passage 16 (Figure 4C).

from 57.13 to 73.90. As expected, the SNR of passage 45 (72.48) was consistent with that of passage 16 (Figure 4C).

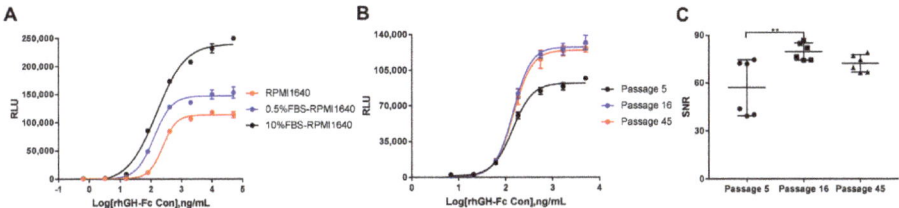

Figure 4. The robustness of the responsive HEK293-Luc cell line to rhGH-Fc. (**A**) The responsiveness curves of HEK293-Luc cells responded to rhGH-Fc in 0.5% and 10% FBS RPMI 1640 assay media. (**B**) The responsiveness of HEK293-Luc cells responded to rhGH-Fc at passage 5, passage 16, and passage 45. (**C**) The SNRs of HEK293-Luc cells responded to rhGH-Fc at passage 5, passage 16, and passage 45. ** $p < 0.01$.

4. Discussion

It is noteworthy that several rhGH-Fc types are at different stages of clinical and preclinical development [7,16]. The bioactivity of rhGH-Fc plays a role in quality control and is required to establish a national or international bioassay. In the published literature, a few studies focused on the proliferation assay and the hGHR-LUR reporter system in some cell lines, e.g., HEK293, Ba/F3 and CHO, but the analytical method has not previously been fully validated. Moreover, transient transfection is not suitable for determination the bioactivity of therapeutic rhGH-Fc [18,21–23,27]. Here, we developed and validated an animal-free bioassay based on RGA for the determination of the bioactivity of rhGH-Fc. It is the first time that the validity of this bioassay to determine the bioactivity of rhGH-Fc has been stringently validated.

The RGAs consist of a reporter gene controlled by a specific regulatory response element and receptors expressed on the cell surface. In this study, the reporter gene system was indicated by firefly luciferase and the SG response element associated with the JAK2-STAT5 signaling pathway [4,18,28,29]. The binding of GH to its receptors can activate JAK2, stimulate tyrosyl phosphorylation of both GHR and JAK2, and initiate signaling pathways. The STAT5 pathway is important to allow GH to exert physiological effects. Phosphorylated STAT5, a major molecule in the STAT5 signaling pathway, translocates to the nucleus, where it binds to the SG response element in the promoter, including SIE and GAS, activates the expression of the downstream gene, and induces cell differentiation and proliferation [2,4,30,31].

During the development phase, the parameters listed in Section 3.2 were optimized based on the SNR and EC$_{50}$ using traditional "single-factor-at-a-time" experiments. Then, a comprehensive validation method was carried out to assess the validity of the bioassay according to the ICH Q2 (R1) guidelines and Chinese Pharmacopoeia. Specificity is considered to be the most important parameter, and it was assessed using both Fc-fusion proteins and degraded compounds of rhGH-Fc. The rhGH-Fc high specificity appeared to be remarkable, specifically for hGHR and SG. Likewise, the bioassay showed high specificity in discriminating the bioactivity spectra of rhGH-Fc.

For the validation of cell-based bioassay, there are no acceptance criteria for precision and accuracy in the ICH Q2 (R1) guidelines. Hence, in our cell-based bioassay, there are 15% to 20% CV as the acceptance criteria, and recommended by the Q2(R3)/FDA Bioanalytical Workshop [25]. In our study, all of the CV values (intra-day and inter-day), including the precision, accuracy, and linearity, were less than 9.00%, which suggests that this method is acceptable for the bioactivity determination of rhGH-Fc. Moreover, the bioactivity determined by two analysts suggested that the bioassay was characterized by excellent intermediate precision. It is expected the bioassay was unaffected by some deliberate changes in method parameters. For one thing, parallel EC$_{50}$ and SNR responses of HEK293-Luc to rhGH-Fc were observed in the two assays media, one thing DMEM supplemented with 20 mM HEPES and the commercial RPMI 1640 with HEPES (data not shown). Despite its easy-to-use nature, we chose the RPMI 1640 without additional HEPES as the assay media. Additionally, the FBS-free RPMI 1640 media and the commercial RPMI 1640 with HEPES (data not shown). Due to use nature, we chose with an identical SNR and decreasing background luciferase values was optimal. The responsiveness of the HEK292-Luc cell line to rhGH-Fc was considered stable between passage 16 and passage 45. It is noteworthy that there is still the risk of potential contaminants of SG inhibitors or activators in the samples or media, although FBS was excluded from the assay media during the bioassay procedure. Generally, any contaminant in media only affects the background signals, but has little influence on the dose–response curve. As for the samples, which are always final products, their purities and impurities should be qualified to reduce the risk of contaminants.

Taken together, a simple and time-saving bioassay was developed for the determination of rhGH-Fc bioactivity. The proposed bioassay was proven to be precise, reproducible, and robust. It has the potential to be used for the routine analysis of bioactivity during the research, development, and manufacture of therapeutic rhGH-Fc, probably for other LAGH products. Further, the bioassay could probably be used as an alternative bioassay to the traditional in vivo rat weight gain assays; however, an agreement study of the two methods to determining the GH bioactivity is required.

Author Contributions: Conceptualization, J.W.; Data curation, W.Y. and X.S.; Funding acquisition, X.S.; Investigation, X.Q.; Methodology, W.Y., L.Y., W.F., Y.L. and C.R.; Project administration, L.Y., J.W. and C.R.; Resources, W.F.; Software, X.S.; Supervision, W.Y., L.Y., J.W. and C.R.; Validation, W.Y., L.Y., L.L. and C.R.; Writing—Review & Editing, L.Y., J.W. and C.R.

Funding: This work was supported by MOST of China (Grant number 2018ZX09101001); and the Improvement of drug quality standards of the Chinese Pharmacopoeia Commission (Grant number 2018S001).

Conflicts of Interest: The authors declare no conflict of interest.

References

1. Aghili, Z.; Zarkesh-Esfahani, S. Taguchi Experimental Design for Optimization of Recombinant Human Growth Hormone Production in CHO Cell Lines and Comparing its Biological Activity with Prokaryotic Growth Hormone. *Drug Res.* **2017**, *68*, 80–88. [CrossRef]
2. Chia, D.J. Minireview: Mechanisms of growth hormone-mediated gene regulation. *Mol. Endocrinol.* **2014**, *28*, 1012–1025. [CrossRef] [PubMed]
3. Herrington, J.; Carter-Su, C. Signaling pathways activated by the growth hormone receptor. *Trends Endocrinol. Metab. TEM* **2001**, *12*, 252–257. [CrossRef]
4. Carter-Su, C.; Schwartz, J.; Argetsinger, L.S. Growth hormone signaling pathways. *Growth Horm. IGF Res.* **2016**, *28*, 11–15. [CrossRef] [PubMed]
5. Yuen, K.C.; Amin, R. Developments in administration of growth hormone treatment: Focus on Norditropin(R) Flexpro(R). *Patient Prefer. Adherence* **2011**, *5*, 117–124. [CrossRef]
6. Christiansen, J.S.; Backeljauw, P.F.; Bidlingmaier, M.; Biller, B.M.; Boguszewski, M.C.; Casanueva, F.F.; Chanson, P.; Chatelain, P.; Choong, C.S.; Clemmons, D.R.; et al. Growth Hormone Research Society perspective on the development of long-acting growth hormone preparations. *Eur. J. Endocrinol.* **2016**, *174*, C1–C8. [CrossRef]
7. Saenger, P.H.; Mejia-Corletto, J. Long-Acting Growth Hormone: An Update. *Endocr. Dev.* **2016**, *30*, 79–97.
8. Cox, G.N.; Rosendahl, M.S.; Chlipala, E.A.; Smith, D.J.; Carlson, S.J.; Doherty, D.H. A long-acting, mono-PEGylated human growth hormone analog is a potent stimulator of weight gain and bone growth in hypophysectomized rats. *Endocrinology* **2007**, *148*, 1590–1597. [CrossRef] [PubMed]
9. Guan, Y.; He, F.; Wu, J.; Zhao, L.; Wang, X.; Huang, L.; Zeng, G.; Ren, B.; Chen, J.; Liao, X.; et al. A long-acting pegylated recombinant human growth hormone (Jintrolong((R))) in healthy adult subjects: Two single-dose trials evaluating safety, tolerability and pharmacokinetics. *J. Clin. Pharm. Ther.* **2018**, *43*, 640–646. [CrossRef]
10. Kim, S.J.; Kwak, H.H.; Cho, S.Y.; Sohn, Y.B.; Park, S.W.; Huh, R.; Kim, J.; Ko, A.R.; Jin, D.K. Pharmacokinetics, Pharmacodynamics, and Efficacy of a Novel Long-Acting Human Growth Hormone: Fc Fusion Protein. *Mol. Pharm.* **2015**, *12*, 3759–3765. [CrossRef]
11. Ku, C.R.; Brue, T.; Schilbach, K.; Ignatenko, S.; Magony, S.; Chung, Y.S.; Kim, B.J.; Hur, K.Y.; Kang, H.C.; Kim, J.H.; et al. Long-acting FC-fusion rhGH (GX-H9) shows potential for up to twice-monthly administration in GH-deficient adults. *Eur. J. Endocrinol.* **2018**, *179*, 169–179. [CrossRef] [PubMed]

12. Luo, X.; Hou, L.; Liang, L.; Dong, G.; Shen, S.; Zhao, Z.; Gong, C.X.; Li, Y.; Du, M.L.; Su, Z.; et al. Long-acting PEGylated recombinant human growth hormone (Jintrolong) for children with growth hormone deficiency: Phase II and phase III multicenter, randomized studies. *Eur. J. Endocrinol.* **2017**, *177*, 195–205. [CrossRef] [PubMed]
13. Wu, M.; Liu, W.; Yang, G.; Yu, D.; Lin, D.; Sun, H.; Chen, S. Engineering of a Pichia pastoris expression system for high-level secretion of HSA/GH fusion protein. *Appl. Biochem. Biotechnol.* **2014**, *172*, 2400–2411. [CrossRef] [PubMed]
14. Kuo, T.T.; Baker, K.; Yoshida, M.; Qiao, S.W.; Aveson, V.G.; Lencer, W.I.; Blumberg, R.S. Neonatal Fc receptor: From immunity to therapeutics. *J. Clin. Immunol.* **2010**, *30*, 777–789. [CrossRef]
15. Beck, A.; Reichert, J.M. Therapeutic Fc-fusion proteins and peptides as successful alternatives to antibodies. *mAbs* **2014**, *3*, 415–416. [CrossRef]
16. Yuen, K.C.J.; Miller, B.S.; Biller, B.M.K. The current state of long-acting growth hormone preparations for growth hormone therapy. *Curr. Opin. Endocrinol. Diabetes Obes.* **2018**, *25*, 267–273. [CrossRef]
17. Kramer, W.G.; Jaron-Mendelson, M.; Koren, R.; Hershkovitz, O.; Hart, G. Pharmacokinetics, Pharmacodynamics, and Safety of a Long-Acting Human Growth Hormone (MOD-4023) in Healthy Japanese and Caucasian Adults. *Clin. Pharmacol. Drug Dev.* **2018**, *7*, 554–563. [CrossRef]
18. Sakatani, T.; Kaji, H.; Takahashi, Y.; Iida, K.; Okimura, Y.; Chihara, K. Lactogenic hormone responsive element reporter gene activation assay for human growth hormone. *Growth Horm. IGF Res.* **2003**, *13*, 275–281. [CrossRef]
19. Yang, Y.; Zhou, Y.; Yu, L.; Li, X.; Shi, X.; Qin, X.; Rao, C.; Wang, J. A novel reporter gene assay for recombinant human erythropoietin (rHuEPO) pharmaceutical products. *J. Pharm. Biomed. Anal.* **2014**, *100*, 316–321. [CrossRef]
20. Yao, W.; Guo, Y.; Qin, X.; Yu, X.; Shi, X.; Liu, L.; Zhou, Y.; Hu, J.; Rao, C.; Wang, J. Bioactivity Determination of a Therapeutic Recombinant Human Keratinocyte Growth Factor by a Validated Cell-based Bioassay. *Molecules* **2019**, *24*, 699. [CrossRef]
21. Ishikawa, M.; Nimura, A.; Horikawa, R.; Katsumata, N.; Arisaka, O.; Wada, M.; Honjo, M.; Tanaka, T. A novel specific bioassay for serum human growth hormone. *J. Clin. Endocrinol. Metab.* **2000**, *85*, 4274–4279. [CrossRef]
22. Ross, R.J.; Leung, K.C.; Maamra, M.; Bennett, W.; Doyle, N.; Waters, M.J.; Ho, K.K. Binding and functional studies with the growth hormone receptor antagonist, B2036-PEG (pegvisomant), reveal effects of pegylation and evidence that it binds to a receptor dimer. *J. Clin. Endocrinol. Metab.* **2001**, *86*, 1716–1723.
23. Wilkinson, I.R.; Ferrandis, E.; Artymiuk, P.J.; Teillot, M.; Soulard, C.; Touvay, C.; Pradhananga, S.L.; Justice, S.; Wu, Z.; Leung, K.C.; et al. A ligand-receptor fusion of growth hormone forms a dimer and is a potent long-acting agonist. *Nat. Med.* **2007**, *13*, 1108–1113. [CrossRef]
24. ICH. *International Conference on Harmonization (ICH) Guidelines ICH Q2(R1), Validation of Analytical Procedures: Text and Methodology*; ICH: Geneva, Switzerland, 2005.
25. Chinese Pharmacopoeia Commission. *Chinese Pharmacopoeia*, 4th ed.; People's Medical Publishing House: Beijing, China, 2015.
26. Hawe, A.; Wiggenhorn, M.; van de Weert, M.; Garbe, J.H.O.; Mahler, H.-C.; Jiskoot, W. Forced Degradation of Therapeutic Proteins. *J. Pharm. Sci.* **2012**, *101*, 895–913. [CrossRef]
27. Maimaiti, M.; Tanahashi, Y.; Mohri, Z.; Fujieda, K. Development of a bioassay system for human growth hormone determination with close correlation to immunoassay. *J. Clin. Lab. Anal.* **2012**, *26*, 328–335. [CrossRef]
28. Karbasian, M.; Kouchakzadeh, H.; Anamaghi, P.N.; Sefidbakht, Y. Design, development and evaluation of PEGylated rhGH with preserving its bioactivity at highest level after modification. *Int. J. Pharm.* **2018**, *557*, 9–17. [CrossRef]
29. New, D.C.; Miller-Martini, D.M.; Wong, Y.H. Reporter gene assays and their applications to bioassays of natural products. *Phytotherapy Res.* **2003**, *17*, 439–448. [CrossRef]
30. Lichanska, A.M.; Waters, M.J. New insights into growth hormone receptor function and clinical implications. *Horm. Res.* **2008**, *69*, 138–145. [CrossRef]
31. Carter-Su, C.; Rui, L.; Herrington, J. Role of the tyrosine kinase JAK2 in signal transduction by growth hormone. *Pediatr. Nephrol.* **2000**, *14*, 550–557. [CrossRef]

32. Herrington, J.; Smit, L.S.; Schwartz, J.; Carter-Su, C. The role of STAT proteins in growth hormone signaling. *Oncogene* **2000**, *19*, 2585–2597. [CrossRef]
33. Hackett, R.H.; Wang, Y.D.; Larner, A.C. Mapping of the cytoplasmic domain of the human growth hormone receptor required for the activation of Jak2 and Stat proteins. *J. Boil. Chem.* **1995**, *270*, 21326–21330. [CrossRef]
34. Argetsinger, L.S.; Campbell, G.S.; Yang, X.; Witthuhn, B.A.; Silvennoinen, O.; Ihle, J.N.; Carter-Su, C. Identification of JAK2 as a growth hormone receptor-associated tyrosine kinase. *Cell* **1993**, *74*, 237–244. [CrossRef]
35. Viswanathan, C.T.; Bansal, S.; Booth, B.; DeStefano, A.J.; Rose, M.J.; Sailstad, J.; Shah, V.P.; Skelly, J.P.; Swann, P.G.; Weiner, R. Quantitative bioanalytical methods validation and implementation: Best practices for chromatographic and ligand binding assays. *Pharm. Res.* **2007**, *24*, 1962–1973. [CrossRef]

Sample Availability: Samples of the compounds rhGH-Fc are available from the authors.

 © 2019 by the authors. Licensee MDPI, Basel, Switzerland. This article is an open access article distributed under the terms and conditions of the Creative Commons Attribution (CC BY) license (http://creativecommons.org/licenses/by/4.0/).

Article

Pharmacokinetics and Tissue Distribution of Alnustone in Rats after Intravenous Administration by Liquid Chromatography-Mass Spectrometry

Yang Song [1,†], Yu Zhou [2,†], Xiao-Ting Yan [1], Jing-Bo Bi [1], Xin Qiu [1], Yu Bian [1], Ke-Fei Wang [1], Yuan Zhang [2] and Xue-Song Feng [1,*]

1 School of Pharmacy, China Medical University, Shenyang 110122, China
2 Department of Pharmacy, National Cancer Center/National Clinical Research Center for Cancer, Chinese Academy of Medical Sciences and Peking Union Medical College, Beijing 100021, China
* Correspondence: xsfeng@cmu.edu.cn; Tel./Fax: +86-024-31939448
† The authors contributed equally to this work.

Academic Editors: Constantinos K. Zacharis and Aikaterini Markopoulou
Received: 8 August 2019; Accepted: 31 August 2019; Published: 2 September 2019

Abstract: Alnustone, a nonphenolic diarylheptanoid, first isolated from *Alnus pendula* (Betulaceae), has recently received a great deal of attention due to its various beneficial pharmacological effects. However, its pharmacokinetic profile in vivo remains unclear. The purpose of this study is to establish a fast and sensitive quantification method of alnustone using liquid chromatography tandem mass spectrometry (LC-MS/MS) and evaluate the pharmacokinetic and tissue distribution profiles of alnustone in rats. The sample was precipitated with acetonitrile with 0.5% formic acid and separated on BEH C_{18} Column. The mobile phase was composed of 0.1% formic acid in water and methanol at a flow rate of 0.3 mL/min. Alnustone and the internal standard (caffeine) were quantitatively monitored with precursor-to-product ion transitions of m/z 262.9→105.2 and m/z 195.2→138.0, respectively. The calibration curve for alnustone was linear from 1 to 2000 ng/mL. The intra- and inter-day assay precision (*RSD*) ranged from 1.1–9.0 % to 3.3–8.6%, respectively and the intra- and inter-day assay accuracy (*RE*) was between −8.2–9.7% and −10.3–9.9%, respectively. The validated method was successfully applied to the pharmacokinetic studies of alnustone in rats. After single-dose intravenous administration of alnustone (5 mg/kg), the mean peak plasma concentration (C_{max}) value was 7066.36 ± 820.62 ng/mL, and the mean area under the concentration-time curve (AUC_{0-t}) value was 6009.79 ± 567.30 ng/mL·h. Our results demonstrated that the residence time of alnustone in vivo was not long and it eliminated quickly from the rat plasma. Meanwhile, the drug is mainly distributed in tissues with large blood flow, and the lung and liver might be the target organs for alnustone efficacy. The study will provide information for further application of alnustone.

Keywords: pharmacokinetics; tissue distribution; alnustone; rats; LC-MS/MS

1. Introduction

Alnustone, a nonphenolic diarylheptanoid with a typical chemical structure of an aryl-C7-aryl skeleton, was first isolated from the male flower of *Alnus pendula* (Betulaceae) [1–3]. Then, it was also found in the seeds of *Alpinia katsumadai* Hayata (Zingiberaceae) [4–10], the rhizomes of *Curcuma xanthorrhiza* Roxb [11,12] and *Curcuma comosa* Roxb (Zingiberaceae) [13]. It is reported that diphenylheptanes have a wide range of pharmacological activities, such as anti-inflammatory, hepatoprotective, antioxidant and anti-tumor effects [14–16]. Diphenylhexane natural drugs account for 5% of the market share of neurosinosidase inhibitors [17]. Alnustone reportedly exhibits a variety of activities, including antihepatotoxic [18], anti-inflammatory [11], antibacterial [4], antiemetic [5,8,9] and weak estrogenic [13]. Recently, Grienke et al. revealed that alnustone showed neuraminidase inhibitory activity, and it was concluded that the compound may be employed as an antiviral agent [6]. Additionally, the chemical compounds from *Alpinia katsumadai* Hayata seeds have been evaluated for their antitumor activities in vitro. Among the isolated compounds, alnustone was found to exhibit significant antitumor activity against the Bel-7402 (human hepatocellular carcinoma cells) and LO-2 (human normal liver cells) cell lines [19]. As a good drug candidate, several methods have been developed to prepare alnustone, including the purification from plants [16] and synthesis through an organic method [20,21].

Although the bioactivities have been investigated extensively, a few analytical methods have been reported. Currently, there are only three articles concerning the analytical methods for the determination of alnustone in natural medical plants or Chinese patent medicine, including HPLC [22,23], semi-preparative liquid chromatography [24]. As far as the authors know, the pharmacokinetics of this compound still remains unknown. It is generally accepted that the study of pharmacokinetics and tissue distribution plays an important role in the drug development because it helps to predict and explain the various issues associated with drug efficacy and toxicity [25,26]. Therefore, it is necessary to establish an effective method to investigate the pharmacokinetic characteristics of alnustone so as to better understand its mechanism of action.

In the present study, a liquid chromatography tandem mass spectrometry (LC-MS/MS) method was developed and validated for the determination of alnustone in rat plasma. The pharmacokinetic behavior and tissue distribution of the intravenous injection of alnustone in rats was subsequently investigated by this method. To the authors' knowledge, this is the first report on the pharmacokinetics of alnustone.

2. Results and Discussion

2.1. Method Establishment

2.1.1. Optimization of LC–MS/MS Conditions

Since the rat plasma contains complex endogenous components, it is necessary to establish a sensitive, rapid and effective method to quantitatively determine the concentration of nanogram alnustone with caffeine (IS) in rat plasma. In this experiment, the electrospray ionization (ESI) source in the positive and negative ion mode was compared. The results showed that the targeted substances can be observed to be more stable and more responsive in the positive ion mode. In order to improve the specificity of the detection method, MS/MS ion transition was monitored in multiple-reaction monitoring (MRM) mode. For each analyte, a precursor ion and two MRM transitions were established, monitoring the more abundant product ion (quantifier ion) for the quantification and the less abundant product ion (qualifier ion) for verification. Figure 1 showed the product ion mass spectrum of alustone and IS with their chemical structures and the chemical bond breaking positions. After giving certain collision energy, the richest and the most stable product ions for alnustone and IS were monitored at m/z 105.2 and m/z 138.0, respectively. The MRM transitions for alnustone and IS were m/z 262.9→105.2 and m/z 195.2→138.0, which were employed for quantitative analysis. The qualifier ions for alnustone and IS were set at m/z 133.1 and m/z 110.1, respectively. The parameters were improved by the maximum intensity observed for product ions, including the ionspray voltage, the collision cell exit potential, the entrance potential, the declustering potential (DP) and the collision energy (CE). Among them, the DP of alnustone and IS were set at 91 V and 92 V; the CE were 18 eV and 30 eV, respectively.

The optimization of chromatographic conditions included the selection of a suitable mobile phase and chromatographic column. The results showed that methanol as the organic phase gave a better peak shape and lower background noise for the analytes than acetonitrile. In addition, in order to improve the ion response and the peak shape of analytes, this study selected and tested two types of ion additives, 5 mmol ammonium acetate and 0.1% fomic acid. The results showed that although the former significantly improved the peak shape, the excessive acid induced-ion suppression decreased the signal of the tested compounds, while the latter was acceptable in improving the peak shape and signal intensity. In addition, the gradient eluting mode was optimized to make the tested compounds completely separated and have good peak shapes, and the interference between the analyte and the IS was avoided. The modified gradient elution conditions were proved to meet the analytical requirements in terms of column equilibrium and carry over. When the flow rate of the mobile phase was 0.3 mL/min, ACQUITY UPLC BEH C_{18} column was used, and the analysis time of a single sample was less than 6.5 min.

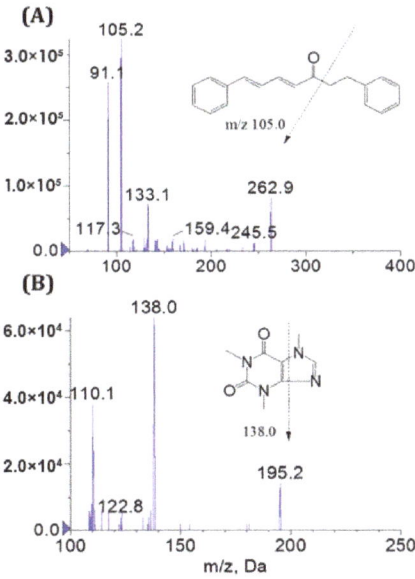

Figure 1. Full-scan product ion spectra of [M+H]$^+$ of alnustone (A) and caffeine (IS, B).

2.1.2. Optimization of Pretreatment Method

To optimize the pretreatment method of the simulated biological samples, a simple and rapid protein precipitation method was attempted as the purification and enrichment method of biological samples. The extraction recovery and matrix effect of alnustone and IS with acetonitrile or methanol as the precipitation reagent were investigated. Acetonitrile was selected as the precipitator due to the higher extraction efficiency and better repeatability. Moreover, the extraction efficiency increased when the acetonitrile contained 0.5% formic acid. The different acetonitrile plasma volume ratios (3:1, 4:1 and 5:1, v:v) were determined. At this ratio of 3:1, the plasma protein turned out to be completely precipitated. In conclusion, the direct protein precipitation method was simple and effective, with high extraction recovery and small background interference.

2.1.3. The Selection of Internal Standard

It is well known that the use of internal standards can improve the accuracy and precision of a quantitative analysis and the stability of the method. Generally speaking, there are two types of internal standards, namely, the structural analogues and the isotope labeling (Sple IS). Although SIL IS has a good performance, it is expensive and difficult to get. In this case, the structural analogue is usually applied as the internal standard. In this study, based on the existing experimental conditions, none of the precipitation reagents were investigated. Available Therefore, several structural analogues were tested as the internal standard, including caffeine, phenoxetine and paracetamol. Among them, caffeine had similar chromatographic behavior and ionization efficiency with the analyte alnustone,

2.2. Method Validation

2.2.1. Selectivity and Carry-Over

No interference peaks were observed at the retention time of alnustone and IS in the chromatograms of six batches of the blank plasma samples. Figure 2 illustrated representative chromatograms of the blank samples, the blank samples spiked with alnustone at the lower limit of the quantification (LLOQ) and IS, and the samples after alnustone administration. The response of the blank samples was compared with the LLOQ samples. The results suggested that there was no obvious endogenous interference in the determination of alnustone and IS.

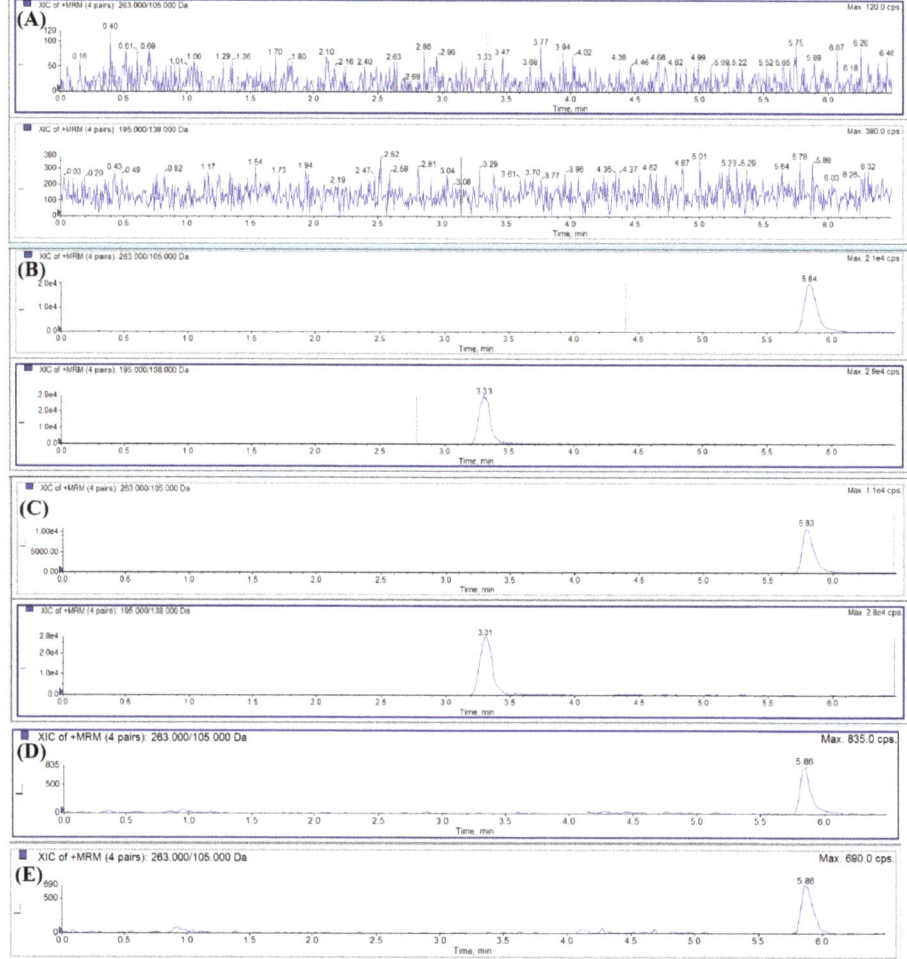

Figure 2. Representative chromatograms of: (A) alnustone (5.84 min) and IS (3.33 min) obtained by the extraction of blank plasma; (B) blank plasma spiked with alnustone and IS; (C) plasma sample from a rat 2h after an intravenous administration of alnustone (5 mg/kg) to rats; (D) blank liver tissue homogenates spiked with alnustone at LLOQ; (E) blank plasma spiked with alnustone at LLOQ.

2.2.2. Linearity

For blood samples and tissue samples, the linear calibration curves were calculated by a linear regression model. The linear equations of the calibration curves, the linear regression coefficients and

In the chromatograms of blank samples injected immediately after the HQC samples were injected, no peak was observed at the retention time of alnustone and IS, indicating that the carry-over effect in the sample analysis could be ignored.

2.2.2. Linearity

For blood samples and tissue samples, the linear calibration curves were calculated by a linear regression model. The linear equations of the calibration curves, the linear regression coefficients and the linear ranges were given in Table 1. Each coefficient of determination (r) in all validation batches were all greater than 0.99.

Table 1. Calibration curves, correlation coefficients, linear ranges and lower limit of the quantification (LLOQ) of alnustone in different biological matrices.

Samples	Calibration Curves	Correlation Coefficients (r)	SES *	SEI #	Linear Ranges (ng/mL)	LLOQs (ng/mL)
Plasma	Y = 0.0256 + 0.00838x	0.996	5.5×10^{-5}	0.13	1–2000	1
Intestine	Y = 0.022 + 0.00302x	0.991	6.7×10^{-5}	0.12	1–2000	1
Heart	Y = 0.0174 + 0.00197x	0.992	1×10^{-4}	0.12	1–2000	1
Liver	Y = 0.063 + 0.00146x	0.993	9.5×10^{-5}	0.10	1–2000	1
Spleen	Y = 0.00038 + 0.00414x	0.992	1.2×10^{-4}	0.18	1–2000	1
Lung	Y = 0.0138 + 0.002x	0.995	1.2×10^{-4}	0.09	1–2000	1
Kidney	Y = 0.0359 + 0.00207x	0.994	8.9×10^{-5}	0.15	1–2000	1
Stomach	Y = −0.00332 + 0.00194x	0.996	6.2×10^{-5}	0.22	1–2000	1
Brain	Y = −0.0458 + 0.00287x	0.995	1.4×10^{-4}	0.17	1–2000	1

* SES—Standard error of the slope ($n = 6$). # SEI—Standard error of the intercept ($n = 6$).

2.2.3. Accuracy and Precision

Table 2. showed the results of intra- and inter-day accuracy and the precision of alnustone in the biological matrices of rats. The *RE* values of intra- and inter-day accuracies ranged from −10.3% to 9.9%, and the *RSD* values of the precision assays were all not more than 9.0%, indicating that the precision and accuracy results of this method were within the acceptable criteria, proving that this method was reliable and reproducible to quantitatively analyze alnustone in rat biological samples.

Table 2. Intra- and inter-day accuracy and precision of alnustone in the plasma and tissue homogenates of rats ($n = 5$).

Samples	QC Conc. (ng/mL)	Intra-Day		Inter-Day	
		Precision (RSD, %)	Accurary (mean%)	Precision (RSD, %)	Accurary (mean%)
Plasma	1	7.2	8.8	7.5	6.9
	5	1.1	9.7	5.0	9.9
	100	6.3	−6.5	4.3	−6.2
	1600	5.2	−8.2	8.1	6.9
Heart	1	8.1	−9.7	6.6	−10.7
	5	6.4	8.0	5.7	−10.3
	100	3.6	3.7	7.5	2.7
	1600	8.0	3.7	4.3	3.1
Liver	1	5.7	4.3	5.8	5.0
	5	4.3	7.7	4.1	−6.6
	100	8.9	9.7	5.6	0.6
	1600	9.0	−5.9	6.1	6.6

Table 2. Cont.

Samples	QC Conc. (ng/mL)	Intra-Day		Inter-Day	
		Precision (RSD, %)	Accurary (mean%)	Precision (RSD, %)	Accurary (mean%)
Spleen	1	9.0	−10.3	7.6	−6.4
	5	7.0	-3.6	5.0	−1.3
	100	7.0	2.2	3.6	5.7
	1600	1.9	8.0	4.1	−6.8
Lung	1	6.6	−8.3	5.8	−5.1
	5	5.7	−6.2	5.3	−1.5
	100	7.5	0.6	6.5	8.1
	1600	5.1	−8.1	6.7	4.3
Kidney	1	4.9	11.2	5.7	10.0
	5	4.2	−6.0	4.4	6.1
	100	6.6	8.0	6.6	1.2
	1600	3.5	−7.2	8.4	−5.3
Brain	1	5.7	8.6	6.1	9.3
	5	7.9	1.3	3.3	9.6
	100	5.9	5.5	8.6	−3.4
	1600	5.9	−4.5	3.3	1.4
Intestine	1	4.9	−7.9	6.8	−4.1
	5	4.1	−1.8	4.3	−6.1
	100	8.0	4.2	5.9	2.6
	1600	7.5	−2.1	6.8	−1.8
Stomach	1	6.2	−8.7	5.3	−5.2
	5	4.9	3.5	3.5	−7.69
	100	2.8	−3.1	7.9	0.8
	1600	7.0	2.8	6.4	−6.3

2.2.4. Recovery and Matrix Effect

As shown in Table 3, the extraction recoveries were within the range of 86.3–110.4%, indicating that the optimized pretreatment method of direct protein precipitation could offer good extraction efficiency for alnustone in these matrices. The matrix effects of alnustone ranged from 89.5% to 114.4%, showing that the matrix had a little co-eluting endogenous substance that could influence the ionization of alnustone.

Table 3. The matrix effect and recovery of alnustone in different samples (n = 5).

Samples	QC Conc. (ng/mL)	Matrix Effect		Recovery	
		Mean ± SD (%)	RSD (%)	Mean ± SD (%)	RSD (%)
Plasma	1	113.7 ± 6.9	5.2	83.5 ± 5.6	5.4
	5	105.4 ± 4.3	4.1	88.7 ±4.4	5.0
	100	91.3 ± 9.6	10.5	95.7 ± 2.2	2.3
	1600	89.5 ± 4.3	4.8	95.4 ± 3.1	3.3
Heart	1	102.1 ± 5.2	5.0	105.1 ± 4.2	4.3
	5	93.8 ± 4.3	4.6	88.3 ± 2.6	3.0
	100	98.2 ± 7.6	7.7	102.3 ± 3.8	3.7
	1600	109.1 ± 3.7	3.4	98.7 ± 1.6	1.7
Liver	1	116.9 ± 10.3	8.0	85.7 ± 5.6	4.1
	5	114.4 ± 9.9	8.6	91.1 ± 3.2	3.5
	100	97.7 ± 9.1	9.3	95.8 ± 4.3	4.4
	1600	98.7 ± 4.6	4.6	90.1 ± 4.0	4.4

Samples	QC Conc. (ng/mL)	Matrix Effect		Recovery	
		Mean ± SD (%)	RSD (%)	Mean ± SD (%)	RSD (%)
Spleen	1	113.5 ± 4.9	6.5	102.9 ± 6.0	4.7
	5	106.1 ± 5.4	5.1	106.9 ± 4.0	3.7
	100	109.8 ± 5.6	5.1	91.2 ± 3.9	4.3
	1600	112.9 ± 4.3	3.8	86.7 ± 0.4	0.4
Lung	1	112.6 ± 6.3	7.9	90.3 ± 6.7	4.6
	5	103.6 ± 7.6	7.3	92.8 ± 3.1	3.3
	100	91.1 ± 9.6	10.6	94.7 ± 1.3	1.3
	1600	92.0 ± 9.8	10.6	87.6 ± 2.7	3.1
Kidney	1	80.7 ± 8.0	5.6	87.8 ± 6.1	5.0
	5	93.9 ± 7.4	7.9	90.1 ± 2.0	2.3
	100	113.4 ± 3.6	3.2	91.8 ± 2.0	2.2
	1600	109.9 ± 8.1	7.4	97.4 ± 5.4	5.5
Brain	1	113.4 ± 6.0	6.2	90.3 ± 6.0	5.8
	5	105.7 ± 5.5	5.2	95.7 ± 4.2	4.4
	100	99.2 ± 3.0	3.0	92.4 ± 3.6	3.9
	1600	94.3 ± 3.4	3.6	110.4 ± 3.5	3.2
Intestine	1	89.6 ± 6.7	6.1	90.7 ± 5.7	5.8
	5	92.0 ± 2.0	2.2	91.8 ± 4.3	4.7
	100	107.3 ± 8.2	7.7	93.4 ± 3.5	3.7
	1600	90.7 ± 7.7	8.5	96.7 ± 3.1	3.2
Stomach	1	113.5 ± 6.2	6.2	88.6 ± 5.6	5.0
	5	110.6 ± 2.8	2.5	93.0± 2.3	2.5
	100	89.9 ± 4.6	5.1	86.3 ± 4.1	4.8
	1600	91.2 ± 3.1	3.3	87.6 ± 2.9	3.3

2.2.5. Stability

The results of the stability studies were shown in Table 4, indicating that there was no stability issue occurred.

Table 4. Short-term, post-preparative storage, freeze-thaw and long-term stability of alnustone in different samples (n = 5).

Samples	QC Conc. (ng/mL)	Short-Term (at Room Temperature for 4 h)	Autosampler 4 °C for 24 h	Three Freeze-Thraw Cycles	Storage at −75°C for 30 d
Plasma	1	108.5 ± 7.6	95.5 ± 5.0	114.1 ± 5.1	98.9 ± 5.7
	5	112.7 ± 8.8	93.4 ± 4.3	112.7 ± 4.9	99.2 ± 6.5
	100	99.8 ± 5.6	101.2 ± 7.5	93.2 ± 7.1	107.2 ± 6.7
	1600	104.8 ± 5.2	108.4 ± 5.5	98.6 ± 2.3	94.8 ± 5.0
Heart	1	109.6 ± 6.0	88.7 ± 8.6	103.8 ± 5.3	107.5± 6.3
	5	104.3 ± 4.3	91.8 ± 8.7	109.1 ± 4.1	101.0 ± 5.3
	100	101.0± 3.2	89.0 ± 3.4	104.8 ± 6.6	112.4 ± 4.0
	1600	91.8 ± 5.8	90.3 ± 5.2	95.3 ± 9.0	95.6 ± 5.4
Liver	1	98.8 ± 6.8	93.1 ± 5.7	101.5 ± 10.1	111.5 ± 6.9
	5	95.4 ± 5.8	95.3 ± 3.2	98.8 ± 9.9	106.5 ± 10.1
	100	111.0 ± 4.9	101.3 ± 5.6	108.4 ± 7.8	105.4 ± 7.2
	1600	106.4 ± 6.0	94.8 ± 7.7	98.3 ± 6.3	97.7 ± 6.8
Spleen	1	108.5± 6.4	108.0 ± 6.0	101.2 ± 7.4	102.3 ± 4.9
	5	114.7 ± 5.8	101.8 ± 4.3	106.2 ± 6.9	98.1 ± 3.9
	100	107.5 ± 8.9	92.9 ± 6.5	105.3 ± 3.1	114.3 ± 6.6
	1600	104.4 ± 6.1	102.1 ± 9.3	92.9 ± 7.6	104.0 ± 3.2

Samples	QC Conc. (ng/mL)	Short-Term (at Room Temperature for 4 h)	Autosampler 4 °C for 24 h	Three Freeze-Thaw Cycles	Storage at −75°C for 30 d
Lung	1	89.5 ± 7.6	105.8 ± 7.0	108.2 ± 9.5	103.4 ± 8.7
	5	92.4 ± 6.4	103.4 ± 7.1	105.8 ± 8.7	93.0 ± 9.6
	100	106.4 ± 8.6	98.4 ± 5.4	94.4 ± 2.6	101.0 ± 7.5
	1600	91.2 ± 6.8	100.6 ± 9.7	93.8 ± 4.6	92.3 ± 6.3
Kidney	1	107.8 ± 7.5	116.5 ± 8.0	93.7 ± 7.0	89.8 ± 8.1
	5	110.7 ± 6.6	114.1 ± 7.4	95.7 ± 5.7	88.5 ± 6.4
	100	109.7 ± 8.3	108.0 ± 5.7	90.9 ± 7.0	111.0 ± 7.5
	1600	100.7 ± 7.0	95.9 ± 6.0	112.5 ± 5.2	100.5 ± 10.6
Brain	1	110.6 ± 5.8	109.2 ± 6.7	86.5 ± 6.1	103.8 ± 4.1
	5	109.0 ± 4.4	104.9 ± 7.0	89.2 ± 4.6	108.9 ± 3.8
	100	99.3 ± 7.5	95.4 ± 5.7	98.0 ± 8.1	90.2 ± 7.4
	1600	96.3 ± 7.4	104.2 ± 4.9	110.2 ± 6.8	112.3 ± 8.9
Intestine	1	89.6 ± 8.1	90.0 ± 5.7	113.7 ± 8.7	96.5 ± 5.7
	5	95.8 ± 7.7	90.9 ± 5.0	108.7 ± 9.4	92.6 ± 3.7
	100	95.3 ± 6.7	98.0 ± 4.9	96.9 ± 3.8	114.4 ± 4.9
	1600	105.9 ± 7.1	103.0 ± 8.4	113.6 ± 2.8	107.0 ± 4.3
Stomach	1	95.1 ± 6.8	93.2 ± 8.5	102.2 ± 5.1	107.4 ± 6.2
	5	93.8 ± 5.9	98.8 ± 9.6	93.2 ± 4.3	94.9 ± 7.2
	100	104.4 ± 5.5	104.4 ± 5.5	114.0 ± 4.9	107.0 ± 6.7
	1600	114.9 ± 6.8	107.9 ± 8.7	104.7 ± 5.8	93.8 ± 7.6

2.3. Pharmacokinetic Study and Tissue Distribution

The validated LC-ESI-MS/MS method was successfully applied to a pharmacokinetics and tissue distribution of alnustone after the intravenous administration at a single dose of 5 mg/kg. The intravenous administration dose was generally determined based on 1/20–1/50 of the LD50 value or 1/2–1/3 of the maximum tolerated dose of the reference MTD [27–29]. The LD50 obtained by our previous pharmacological experiment was 200 mg/kg and therefore the selected i.v. dose was 5 mg/kg. Meanwhile, it was reported that the structural analog of alnustone was given to the rat i.v. at a dose of 4.5 mg/kg for a pharmacokinetic study [30]. The dosage of another piece of literature was 5.45 mg/kg for the pharmacological activity study [31]. Above all, the dosage of the drug administration in our pharmacokinetics experiment was determined to be 5 mg/kg. Figure 3 showed the mean plasma concentration-time profile of alnustone after administration. Table 5 summarized the pharmacokinetic parameters based on the non-compartmental method. Upon intravenous administration of 5 mg/kg alnustone, the T_{max} was 2 min and the C_{max} was 7066.36 ± 820.62 ng/mL, indicating that alnustone could be quickly detected in the plasma. The pharmacokinetic results have demonstrated the area under the curve up to 10 h (AUC_{0-t}) and the infinite time ($AUC_{0-\infty}$) of 6009.79 ± 567.30 and 6032.45 ± 472.50 ng/mL·h, respectively. The total mean residence time up to 10 h (MRT_{0-t}) was found to be 1.60 ± 0.22 h. Alnustone had a relatively short $t_{1/2}$ of 1.31 ± 0.19 h, an apparent V_d of 1.57 ± 0.18 L/kg, and CL of 0.83 ± 0.09 L/h/kg. The short elimination half-life of alnustone in vivo after a single administration indicated that alnustone would not tend to accumulate in rat plasma during long-term use.

Table 5. Non-compartmental plasma pharmacokinetic parameters following a single intravenous administration of alnustone (5 mg/kg) to rats ($n = 12$).

Pharmacokinetic Parameters	Dose of i.v. Administration (5 mg/kg)
Cmax (ng/mL)	7066.36 ± 820.62
$t_{1/2}$ (h)	1.31 ± 0.19
AUC_{0-t} (ng/mL·h)	6009.79 ± 567.30
$AUC_{0-\infty}$ (ng/mL·h)	6032.45 ± 472.50
$MRT_{0-\infty}$ (h)	1.60 ± 0.22
CL (L/h/kg)	0.83 ± 0.09
V_d (L/kg)	1.57 ± 0.18

± 567.30 and 6032.45 ± 472.50 ng/mL·h, respectively. The total mean residence time up to 10 h (MRT_{0-t}) was found to be 1.60 ± 0.22 h. Alnustone had a relatively short $t_{1/2}$ of 1.31 ± 0.19 h, an apparent V_d of 1.57 ± 0.18 L/kg, and CL of 0.83 ± 0.09 L/h/kg. The short elimination half-life of alnustone in vivo after a single administration indicated that alnustone would not tend to accumulate in rat plasma during long-term use.

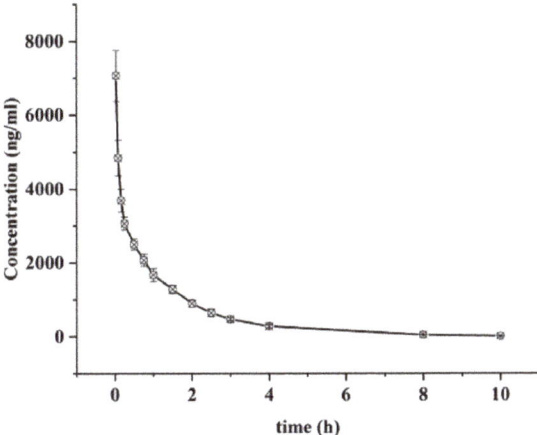

Figure 3. The mean plasma concentration-time curves of alnustone after intravenous administration of alnustone at a dose of 5 mg/kg to rats (n = 12, mean ± SD).

The changing trend of alnustone concentration in different tissues at different time points of 0.5, 1, 2 and 4 h after administration was investigated. Figure 4 revealed that alnustone distributed rapidly and widely in different tissues. In the tissues of brain, intestine, lung, and spleen, the maximum concentration appeared at 1 h, while in heart, kidney, liver and stomach, the maximum concentration of alnustone appeared earlier at 0.5 h to 1h. Among all the tissues, the highest mean concentration was found in the lung (5685.40 ng/g), followed by the liver (3199.49 ng/g), spleen (2077.14 ng/g), stomach (1588.90 ng/g), kidney (1450.43 ng/g), intestine (1099.14 ng/g), heart (900.97 ng/g), and brain (344.27 ng/g). The high level in the lung and liver might be related to the bioaccumulation of alnustone in these two organs. In the liver, an increase of drug exposure was observed at 4 h. Similar situations also occurred in the lung and stomach. The most possible reason was that alnustone might be reabsorbed in these organs. Alustone is fat-soluble, so it might bind to the fat after the first absorption. When the drug reached a certain accumulation in the fat, it would be reabsorbed into the liver, lung and stomach, which have good affinity with the drug. In addition, hepatointestinal circulation may also lead to an increase in the liver at 4 h. Thus, in the liver, lung and stomach, alustone maintained a high concentration and underwent a slow elimination and accumulation. They are highly likely to be the target organs for the efficacy of alnustone, which needs to be studied systematically in the future. In addition, the metabolic clearance rate of alnustone in different tissues was quite different. For example, in the spleen, intestine, and kidney, the elimination rate of alnustone was relatively fast. For 4 h after administration, the concentration of alnustone in these tissues had decreased to less than half of the corresponding maximum concentration. In the other tissues, the elimination rate of alnustone was relatively slow, and the tissue concentrations were either kept at the same level during the measured period or even higher within 4 h after administration. In contrast, due to the rapid elimination and less drug accumulation, alnustone was unlikely to cause drug accumulation and immunological side effects in the spleen, intestine, and kidney. At the same time, the concentration of alnustone in the brain did not change with time, and reached a certain level (567.0 ng/mL, 4 h), suggesting that it could enter the brain through the blood-brain barrier. The tissue distribution experiment provided a reference for the study of the distribution characteristics of drugs in vivo and also provided a theoretical basis for the development of drugs in the later stage.

Table 1. Non-compartmental plasma pharmacokinetic parameters following a single intravenous administration of alnustone (5 mg/kg) to rats (n = 12, mean ± SD).

Pharmacokinetic Parameters	Dose of i.v. Administration (5 mg/kg)
Cmax(ng/mL)	7066.36 ± 820.62
AUC_{0-t} (ng/mL·h)	6009.79 ± 567.30

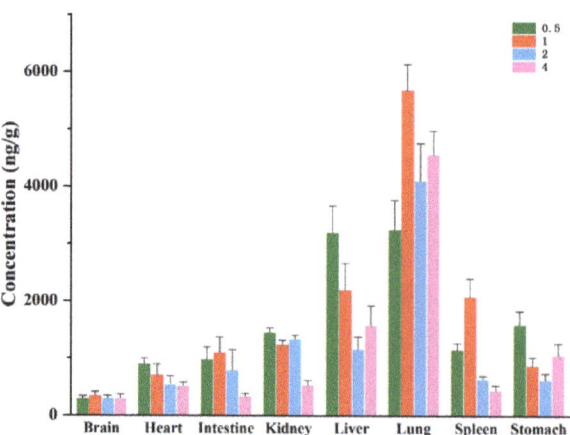

Figure 4. The mean concentration of alnustone in different tissues (ng/g) at 0.5, 1, 2 and 4 h after an intravenous administration of 5 mg/kg alnustone to the rats (Mean ± SD, $n = 5$).

The tissue distribution characteristics of alnustone were determined at 4 h after i.v. administration of alnustone at 5 mg/kg. The tissue to plasma partition coefficients (Kp) shows an upward trend in administration of alnustone at 5 mg/kg. The Kp values of alnustone in iv administration are summarized in Table 6. The highest Kp was observed in the lung (15.84 ± 1.33), followed by that in the liver (5.48 ± 0.66), stomach (3.66 ± 0.27) tissue, and kidney (1.86 ± 0.18). The highest Kp values and concentrations of alnustone in the lung and liver indicated the possible bioaccumulation of alnustone.

Table 6. Plasma to tissue partition coefficients (Kp) of alnustone after i.v. administration of alnustone (5 mg/kg) to the rats (Mean ± SD, $n = 5$).

Time (h)	Heart	Liver	Spleen	Lung	Kidney	Brain	Intestine	Stomach
0.5	0.36 ± 0.06	1.28 ± 0.34	0.46 ± 0.06	1.30 ± 0.09	0.58 ± 0.07	0.11 ± 0.40	0.39 ± 0.07	0.64 ± 0.14
1	0.42 ± 0.10	1.31 ± 0.18	1.24 ± 0.13	3.39 ± 0.39	0.74 ± 0.11	0.21 ± 0.56	0.66 ± 0.12	0.52 ± 0.11
2	0.59 ± 0.06	1.28 ± 0.29	0.70 ± 0.06	4.55 ± 0.61	1.49 ± 0.27	0.33 ± 0.10	0.88 ± 0.19	0.70 ± 0.03
4	1.78 ± 0.24	5.48 ± 0.66	1.54 ± 0.32	15.84 ± 1.33	1.86 ± 0.18	1.03 ± 0.08	1.18 ± 0.20	3.66 ± 0.27

3. Materials and Methods

3.1. Chemicals and Reagents

Alnustone and caffeine (purity over 98%) were provided by Target Molecule Corp. (TargetMol) (Boston, MA, USA). Acetonitrile and methanol, both are LC-MS-grade, were purchased from Merck KGaA Company (Darmstadt, Germany). The formic acid (HCOOH, HPLC-grade) was obtained from Dikma Technologies Inc. (Lake Forest, CA 92630, USA). The purified water was provided using a Millipore Milli-Q system (Millipore, Bedford, MA, USA).

3.2. Instrumentations

The UHPLC-MS/MS method was performed on an Agilent series 1290 UHPLC system (Agilent Technologies, Santa Clara, CA, USA), which was coupled to an AB 3500 triple quadrupole mass spectrometer (AB Sciex, Ontario, ON, Canada) with an electrospray ionization (ESI) source. The separation process was performed on an ACQUITY UPLC BEH C_{18} Column (100 mm × 2.1 mm, 1.7 μm, Agilent Technologies, Santa Clara, CA, USA).

3.3. Animals

Further, thirty two Sprague-Dawley rats (male, 200–220 g body weight) were purchased from the Experimental Animal Research Center, China Medical University, China. The protocol for this study (protocol number # CMU2019194) was approved by the Institutional Animal Care and Use Committee at China Medical University. The study complied with guidelines for the Care and Use of Laboratory Animals (published by the National Institutes of Health, NIH publication no. 85–23).

3.4. Preparation of Calibration Standards and Quality Control Samples

A 10.0 mg of alnustone was dissolved in 10.0 mL of methanol to give the stock solution with the concentration of 1.0 mg/mL. The working solutions were prepared by serial dilution of the stock solution with the initial mobile phase (methanol—0.1% fomic acid water, 20:80, *v/v*). A 10.0 mg of caffeine (IS) was dissolved in 10.0 mL of methanol to give the IS stock solution with the concentration of 1.0 mg/mL. All of the solutions were stored at 4 °C before use.

The calibration standards were prepared by spiking a certain volume of blank plasma or blank tissue homogenates with appropriate amounts of working solutions to yield a final concentration range from 1 to 2000 ng/mL. The effective concentrations of alnustone were 1, 5, 10, 40, 160, 200, 800, 2000 ng/mL for the plasma samples and tissue homogenates samples (heart, liver, spleen, lung, kidney, brain, stomach, intestine). The quality control (QC) samples were prepared in the same manner at three concentration levels (low, mid, high) of 5, 100, 1600 ng/mL for the plasma and tissue homogenates.

3.5. Preparation of Plasma and Tissues Samples

In this study, a direct protein precipitation method was applied to extract alnustone and IS from the biological matrix. An aliquot 100 μL of plasma, 50 μL of IS solution (1 μg/mL) and 200 μL of precipitate agent acetonitrile with 0.5% formic acid was added into a 1.5 mL Eppendorf tube. The mixture was vortex-mixed for 30 s at room temperature, and centrifuged at 12,000× *g* for 10 min. Then, a 200 μL aliquot of the supernatant was carefully removed and transferred to a new 1.5 mL Eppendorf tube and evaporated to dryness at 40 °C under a slight stream of nitrogen. The residuals were reconstituted in 100 μL of methanol—0.1% fomic acid water (20:80, *v/v*) by vortex mixing for 30 s. After centrifuging at 12,000× *g* for 10 min, a 10 μL aliquot of the supernatant was used for the UHPLC-MS/MS analysis.

The rat was sacrificed quickly by decapitation on the ice. The various tissues (heart, liver, spleen, lung, kidney, brain, stomach, intestine) were harvested and rinsed with ice-cold 0.9% NaCl to remove the superficial blood. After being blotted dry with filter paper, each tissue sample was weighed on ice and physiological saline added (1:2, *w/v*) to homogenize. Then, a 100 μL (equivalent to 50 mg) of tissue homogenate was taken and processed using the same method as the plasma samples processing method as shown in the above steps.

3.6. Chromatographic and Mass Conditions

The mobile phase consisted of methanol and 0.1% formic acid water using a gradient dilution at a flow rate of 0.3 mL/min. The column temperature was maintained at 30 °C. Alnustone was quantitatively determined with MRM in the positive ion mode, and nitrogen was used to assist nebulization in the ESI source. The MS parameters were set as follows: The ionspray voltage (IS) 5500 V; nebulizer gas (gas 1) 19 arbitrary units; curtain gas (CUR) 10 arbitrary units; collision cell exit potential (CXP) 7.0 V; entrance potential (EP) 10.0 V. The optimization of the MS transitions using the multiple-reaction monitoring (MRM) mode were accomplished as alnustone *m/z* 262.9→105.2 and IS (caffeine) *m/z* 195.2→138.0, respectively.

3.7. Method Validation

The method was fully validated according to the Bioanalytical Method Validation Guidance for Industry (FDA, 2018) [32].

3.7.1. Selectivity

Selectivity was investigated by comparing chromatograms of blank rat plasma obtained from six individual rats with those of the corresponding standard biosample spiked with alnustone at LLOQ and IS and the actual plasma sample 2h after an intravenous administration of alnustone (5 mg/kg).

3.7.2. Linearity and LLOQ

The calibration curves for alnustone in the plasma or tissue homogenates were generated by plotting the peak area ratios (y) of alnustone to IS versus those nominal concentrations (x) in standard plasma or tissue homogenates using weighted ($1/x^2$) least squares linear regression. The LLOQ was defined as the lowest concentration of the calibration curve at which the precision did not exceed 20%, the accuracy was within ±20% and the signal-to-noise ratio (S/N) was at least 10.

3.7.3. Accuracy and Precision

The intra- and inter-day accuracy and precision were assessed by analyzing QC samples at three levels of alnustone (5, 100, 1600 ng/mL for plasma and tissue homogenates) on three consecutive days with five replicates at each concentration. The accuracy and precision were depicted as the relative error (*RE*) and the relative standard deviation (*RSD*), respectively. The *RE* was within ±15% and the *RSD* could not exceed 15%.

3.7.4. Recovery and Matrix Effect

The recovery was determined by comparing the peak areas of alnustone or IS in processed QC samples with the mean peak areas of alnustone or IS in the spike-after-extraction samples (blank plasma or tissue homogenates extracted then spiked with QC standards). ($n = 5$)

The matrix effect was evaluated by comparing the peak areas of alnustone or IS in the spike-after-extraction samples (blank plasma or tissue homogenates extracted then spiked with QC standards) with the mean peak areas of alnustone or IS dissolved with the mobile phase at high, medium and low levels, respectively. ($n = 5$). It was generally considered that the matrix effect was obvious if the ratio was <85% or >115%.

3.7.5. Stability

The stability tests involved the following four conditions. The short-term stability study was examined by analyzing samples at room temperature for 4 h. The long-term stability study was performed by analyzing samples stored at −75 °C for 1 month. For the freeze-thaw stability study, the samples were analyzed after three freeze/thaw cycles (−75 °C to 25 °C). The post-preparative storage stability was assessed by analyzing the samples left in autosampler vials at 4 °C for 24 h. The stability analysis was performed using three aliquots of each QC samples at three different concentrations. The samples were considered stable if the assay values were within the acceptable limits of accuracy (±15% *RE*) and precision (≤15% *RSD*).

3.8. Drug Administration and Sampling

This study was approved by the Animals Experimental Ethical Committee of China Medical University (Liaoning, China). Alnustone was dissolved in 0.5% (v/v) DMSO saline to give the injection solution at the dose of 5 mg/kg for intravenous administration.

For the pharmacokinetic study, 12 SD rats fasted for 12 h with free access to water prior to administration. After the rats were tail intravenously administrated of alnustone (5 mg/kg), a blood sample (0.4 mL) was collected from the orbital vein into heparinized tubes at appropriate time intervals (2 min, 5 min, 10 min, 15 min, 30 min, 45 min, 1 h, 1.5 h, 2 h, 2.5 h, 3 h, 4 h, 8 h, 10 h). The blank plasma samples were prepared before dosing. All blood samples were immediately centrifuged at 12,000× g for 10 min and stored at −75 °C until analysis.

For the tissue distribution study, 20 SD rats were randomly divided into four groups (5 rats for each time point) and fasted for 12 h with free access to water prior to administration. After the rats were tail intravenously administrated of alnustone (5 mg/kg), the tissue specimens (heart, liver, spleen, lung, kidney, brain, stomach and intestine) were collected at 0.5, 1, 2, and 4 h post-dosing, respectively. The blank tissues were prepared separately using drug-free SD rats. The tissue harvesting and homogenizing method were given in Section 3.5. For the plasma and tissue homogenate sample preparation method, also see Section 3.5.

3.9. Statistical Analysis

The pharmacokinetic parameters were evaluated using DAS 3.2.8 pharmacokinetic program [33].

4. Conclusions

In this study, a rapid, simple and sensitive UHPLC–ESI MS/MS method was established and successfully applied to the study of pharmacokinetics and tissue distribution of alnustone after a single intravenous administration with 5 mg/kg alnustone in rats. The method has been validated for the first time in this paper and proved to be able to detect a low concentration of 1 ng/mL for alnustone with one direct protein precipitation method. The elimination half-life of alnustone was short implying the residence time of alnustone in vivo was not long and it eliminated quickly from the rat plasma. According to the investigation of tissue distribution, alnustone was mainly distributed in the lung and liver tissues within 1 h, and the drug in each tissue increased slightly at 4 h, which indicated that the drug was mainly distributed in the tissues with large blood flow, and the lung and liver might be the target organs for the curative effect of alnustone. The developed and validated LC-MS/MS method and the pharmacokinetic study of alnustone in the present paper might provide helpful information for its future preclinical applications and theoretical basis for its further exploitation.

Author Contributions: The experiments were conceived and designed by X.-S.F. The experiments were equally performed by Y.S. and Yu.Z., Chromatographic analysis of plasma and biosamples were performed by X.-T.Y., J.-B.B., Analysis of pharmacokinetic behavior of alnustone was finished by X.Q., Y.B., K.-F.W. and Yua.Z., All authors read and approved the final manuscript.

Funding: This project was financially supported by Key Program of the Natural Science Foundation of Liaoning Province of China (No. 20170541027); Liaoning planning Program of philosophy and social science (No. L17BGL034); Major subject in education research founded by Chinese Medical Association Medical Education Branch and China Association of Higher Education Medical Education Professional Committee (2018A-N19012); and Scientific Research Project of Department of Education of Liaoning Province (ZF2019036).

Acknowledgments: We are very grateful to Heran Li and Xiaolan Deng for making a major contribution to the revision of the article. The authors are grateful to every person who helped in this work.

Conflicts of Interest: The authors declare no conflicts of interest.

References

1. Suga, T.; Asakawa, Y.; Iwata, N. 1,7-diphenyl-1,3-heptadien-5-ON, einneues keton aus *Alnus pendula* (Betulaceae). *Chem. Ind. (London)* **1971**, *27*, 766–768.
2. Suga, T.; Iwata, N.; Asakawa, Y. Chemical constituents of the male flower of *Alnus pendula* (Betulaceae). *Bull. Chem. Soc. Jpn.* **1972**, *7*, 2058–2060. [CrossRef]
3. Aoki, T.; Ohta, S.; Suga, T. Triterpenoids, diarylheptanoids and their glycosides in the flowers of *Alnus* species. *Phytochemistry* **1990**, *29*, 3611–3614. [CrossRef]
4. Huang, W.Z.; Dai, X.J.; Liu, Y.Q.; Zhang, C.F.; Zhang, M.; Wang, Z.T. Studies on antibacterial activity of flavonoids and diarylheptanoids from *Alpinia katsumadai*. *J. Plant Resour. Environ.* **2006**, *1*, 37–40.
5. Huang, W.Z.; Zhang, C.F.; Zhang, M.; Wang, Z.T. A new biphenylpropanoid from *Alpinia katsumadai*. *J. Chin. Chem. Soc.* **2007**, *6*, 1553–1556. [CrossRef]
6. Grienke, U.; Schmidtke, M.; Kirchmair, J.; Pfarr, K.; Wutzler, P.; Durrwald, R.; Wolber, G.; Liedl, K.R.; Stuppner, H.; Rollinger, J.M. Antiviral potential and molecular insight into neuraminidase inhibiting diarylheptanoids from *Alpinia katsumadai*. *Med. Chem.* **2010**, *2*, 778–786. [CrossRef]

7. Kuroyanagi, M.; Noro, T.; Fukushima, S.; Aiyama, R.; Ikuta, A.; Itokawa, H.; Morita, M. Studies on the Constituents of the Seeds of *Alpinia katsumadai* Hayata. *Chem. Pharm. Bull.* **1983**, *31*, 1544–1550. [CrossRef]
8. Yang, Y.; Kinoshita, K.; Koyama, K.; Takahashi, K.; Tai, T.; Nunoura, Y.; Watanabe, K. Anti-emetic principles of *Alpinia katsumadai* Hayata. *Nat. Prod. Sci.* **1999**, *5*, 20–24.
9. Yang, Y.; Kinoshita, K.; Koyama, K.; Takahashi, K.; Kondo, S.; Watanabe, K. Structure-antiemetic-activity of some diarylheptanoids and their analogues. *Phytomedicine* **2002**, *9*, 146–152. [CrossRef]
10. Nam, J.W.; Seo, E.K. Structural characterization and biological effects of constituents of the seeds of *Alpinia katsumadai* (*Alpina Katsumadai* Seed). *Nat. Prod. Commun.* **2012**, *6*, 795–798. [CrossRef]
11. Claeson, P.; Panthong, A.; Tuchinda, P.; Reutrakul, V.; Kanjanapothi, D.; Taylor, W.C.; Santisuk, T. Three non-phenolic diarylheptanoids with anti-inflammatory activity from *Curcuma xanthorrhiza*. *Planta Med.* **1993**, *5*, 451–454. [CrossRef] [PubMed]
12. Clseson, P.; Pongprayoon, U.; Sematong, T.; Tuchinda, P.; Reutrakul, V.; Soontornsaratune, P.; Taylor, W.C. Non-phenolic linear diarylheptanoids from *Curcuma xanthorrhiza*: A novel type of topical anti-inflammatory agents: Structure activity relation ship. *Planta Med.* **1996**, *62*, 236–240. [CrossRef] [PubMed]
13. Suksamrarn, A.; Ponglikitmongkol, M.; Wongkrajang, K.; Chindaduang, A.; Kittidanairak, S.; Jankam, A.; Yingyongnarongkul, B.E.; Kittipanumat, N.; Chokchaisiri, R.; Khetkam, P. Diarylheptanoids, new phytoestrogens from the rhizomes of *Curcuma comosa*: Isolation, chemical modification and estrogenic activity evaluation. *Bioorg. Med. Chem.* **2008**, *16*, 6891–6902. [CrossRef] [PubMed]
14. Ishida, J.; Kozuka, M.; Tokuda, H.; Nishino, H.; Nagumo, S.; Lee, K.H.; Nagai, M. Chemopreventive potential of cyclic diarylheptanoids. *Bioorg. Med. Chem.* **2002**, *10*, 3361–3365. [CrossRef]
15. Song, E.; Cho, H.; Kim, J.S.; Kim, N.Y.; An, N.H.; Kim, J.A.; Lee, S.H.; Kim, Y.C. Diarylheptanoids with free radical scavenging and hepato protective activity in vitro from *Curcuma longa*. *Planta Med.* **2001**, *67*, 876–877. [CrossRef] [PubMed]
16. Yamazaki, R.; Hatano, H.; Aiyama, R.; Matsuzaki, T.; Hashimoto, S.; Yokokura, T. Diarylheptanoids suppress expression of leukocyte adhesion molecules on human vascular endothelial cells. *Eur. J. Pharmacol.* **2000**, *3*, 375–385. [CrossRef]
17. Grienke, U.; Schmidtke, M.; von Grafenstein, S.; Kirchmair, J.; Liedl, K.R.; Rollinger, J.M. Influenza neuraminidase: A druggable target for natural products. *Nat. Prod. Rep.* **2012**, *1*, 11–36. [CrossRef] [PubMed]
18. Hikino, H.; Kiso, Y.; Kato, N.; Hamada, Y.; Shioiri, T.; Aiyama, R.; Itokawa, H.; Kiuchi, F.; Sankawa, U. Antihepatotoxic actions of gingerols and diarylheptanoids. *J. Ethnopharmacol.* **1985**, *1*, 31–39.
19. Li, Y.Y.; Li, Y.; Wang, C.H.; Chou, G.X.; Wang, Z.T. Chemical constituents from seeds of *Alpinia Katsumadai* Hayata. *Acta Univ. Tradit. Med. Sin. Pharmacol. Shanghai* **2010**, *24*, 72–75.
20. Kucukoglu, K.; Seçinti, H.; Ozgur, A.; Seçen, H.; Tutar, Y. Synthesis, molecular docking, and antitumoral activity of alnustone-like compounds against estrogen receptor alpha-positive human breast cancer. *Turkish J. Chem.* **2015**, *1*, 179–193. [CrossRef]
21. Goksu, S.; Celik, H.; Secen, H. An efficient synthesis of alnustone, a naturally occurring compound. *Turkish J. Chem.* **2002**, *1*, 31–34.
22. Li, Y.Y.; Chou, G.X.; Yang, L.; Wang, Z.T. Qualitative and quantitative methods for *Alpiniae Katsumadai* Semen. *China J. Chin. Mater. Medica* **2010**, *16*, 2091–2094.
23. Sun, H.; He, S.L. Quality control for components in Yangweining capsule based on HPLC wavelength switching method. *Anhui Med. Pharmaceutical J.* **2019**, *3*, 473–476.
24. Wang, C.Y.; Li, A.F.; Sun, A.L.; Liu, R.M. Separation of Active Components in *Alpinia katsumadai* Hayata by Semi-prepartive High Performance Liquid Chromatography. *J. Anhui Agri. Sci.* **2016**, *24*, 127–130.
25. Lakota, E.A.; Bader, J.C.; Rubino, C.M. Ensuring quality pharmacokinetic analyses in antimicrobial drug development programs. *Curr. Opin. Pharmacol.* **2017**, *36*, 139–145. [CrossRef]
26. Zhu, L.; Wu, J.; Zhao, M.; Song, W.; Qi, X.; Wang, Y.; Lu, L.; Liu, Z. Mdr1a plays a crucial role in regulating the analgesic effect and toxicity of aconitine by altering its pharmacokinetic characteristics. *Toxicol. Appl. Pharmacol.* **2017**, *320*, 32–39. [CrossRef] [PubMed]
27. Chen, Q. *Research Methods in Pharmacology of Chinese Materia Medica*, 3rd ed.; People's Medical Publishing House: Beijing, China, 2011.
28. Sharma, D.R.; Sunkaria, A.; Bal, A.; Bhutia, Y.D.; Vijayaraghavan, R.; Flora, S.J.; Gill, K.D. Neurobehavioral impairments, generation of oxidative stress and release of pro-apoptotic factors after chronic exposure to sulphur mustard in mouse brain. *Toxicol. Appl. Pharmacol.* **2009**, *15*, 208–218. [CrossRef]

29. Tao, X.; Cush, J.J.; Garret, M.; Lipsky, P.E. A phase I study of ethyl acetate extract of the chinese antirheumatic herb Tripterygium wilfordii hook F in rheumatoid arthritis. *J. Rheumatol.* **2005**, *10*, 2160–2167.
30. Huang, X.Y.; Duan, Q.Y.; Liu, J.X.; Di, D.L. Determination of a novel diarylheptanoid (Juglanin B) from green walnut husks (Juglans regia L.) in rat plasma by high-performance liquid chromatography. *Biomed. Chromatogr.* **2010**, *3*, 307–311.
31. Su, J.; Sripanidkulchai, K.; Suksamrarn, A.; Hu, Y.; Piyachuturawat, P.; Sripanidkulchai, B. Pharmacokinetics and organ distribution of diarylheptanoid phytoestrogens from Curcuma comosa in rats. *J. Nat. Med.* **2012**, *3*, 468–475. [CrossRef]
32. U.S. Department of Health and Human Services, Food and Drug Administration, Center for Drug Evaluation and Research (CDER), Center for Veterinary Medicine (CVM), Bioanalytical Method Validation Guidance for Industry 2018. Available online: https://www.fda.gov/files/drugs/published/Bioanalytical-Method-Validation-Guidance-for-Industry.pdf (accessed on 30 August 2019).
33. Center of Clinical Drug Research, Shanghai University of Traditional Chinese Medicine, Shanghai, People's Republic of China. Drug Analysis System [CP/DK]. Available online: http://www.drugchina.net/# (accessed on 30 August 2019).

Sample Availability: Samples of the compounds alnustone and caffeine are available from the authors.

© 2019 by the authors. Licensee MDPI, Basel, Switzerland. This article is an open access article distributed under the terms and conditions of the Creative Commons Attribution (CC BY) license (http://creativecommons.org/licenses/by/4.0/).

Article

Pharmacokinetics and Tissue Distribution of Anwuligan in Rats after Intravenous and Intragastric Administration by Liquid Chromatography-Mass Spectrometry

Yang Song [1,†], Yuan Zhang [2,†], Xiao-Yi Duan [1], Dong-Wei Cui [1], Xin Qiu [1], Yu Bian [1], Ke-Fei Wang [1] and Xue-Song Feng [1,*]

1. School of Pharmacy, China Medical University, Shenyang 110122, China; songyanglhyb1998@163.com (Y.S.); dxychn@126.com (X.-Y.D.); cdwhhazj@163.com (D.-W.C.); 13314227291@163.com (X.Q.); bianyu2468346755@126.com (Y.B.); wl5712367805@sina.com (K.-F.W.)
2. Department of Pharmacy, National Cancer Center/National Clinical Research Center for Cancer, Chinese Academy of Medical Sciences and Peking Union Medical College, Beijing 100730, China; 13840149878@163.com
* Correspondence: xsfeng@cmu.edu.cn; Tel.:+86-024-31939448
† These authors contributed equally to this work.

Academic Editors: Constantinos K. Zacharis and Aikaterini Markopoulou
Received: 3 December 2019; Accepted: 19 December 2019; Published: 20 December 2019

Abstract: Anwuligan, a natural 2,3-dibenzylbutane lignan from the nutmeg mace of *Myristica fragans*, has been proved to possess a broad range of pharmacological effects. A rapid, simple, and sensitive liquid chromatography tandem mass spectrometry (LC-MS/MS) method has been established and successfully applied to the study of pharmacokinetics and tissue distribution of anwuligan after intravenous or intragastric administration. Sample preparation was carried out through a liquid-liquid extraction method with ethyl acetate as the extraction reagent. Arctigenin was used as the internal standard (IS). A gradient program was employed with a mobile phase consisting of 0.1% formic acid aqueous solution and acetonitrile. The mass spectrometer was operated in a positive ionization mode with multiple reaction monitoring. The transitions for quantification were m/z 329.0→205.0 for anwuligan and m/z 373.0→137.0 for IS, respectively. Calibration curves were linear over the ranges of 0.5–2000 ng/mL for both plasma samples and tissue samples (r > 0.996). The absolute bioavailability is 16.2%, which represented the existing of the obvious first-pass effect. An enterohepatic circulation was found after the intragastric administration. Anwuligan could be distributed rapidly and widely in different tissues and maintained a high concentration in the liver. The developed and validated LC-MS/MS method and the pharmacokinetic study of anwuligan would provide reference for the future investigation of the preclinical safety of anwuligan as a candidate drug.

Keywords: anwuligan; LC-MS/MS; pharmacokinetics; tissue distribution; rat

1. Introduction

Anwuligan is a natural 2,3-dibenzylbutane lignan from the nutmeg mace of *Myristica fragans* [1,2]. *Myristica fragans* (*M. fragans*) is a tropical evergreen tree native to Indonesia and cultivated in India, Iran, the West Indies and South America. The mace found in *M. fragans* has been reported to have anti-fungal and anti-bacterial activities [3]. Anwuligan has been proved to possess a broad range of pharmacological effects (see Table S1), including anti-bacterial, anti-inflammatory, and anti-cancer activities. Recently, it has been proved to have anti-diabetic, hepatoprotective, and neuroprotective effects. In terms of antibacterial activity, anwuligan could inhibit the growth of various bacteria, including *Bacillus cereus* [4], *Streptoccus* bacteria [5], *Lactobacillus* bacteria [5]; it could also effectively inhibit oral colonizing bacteria and has the potential to remove these bacteria and reduce the plaque formation [6]. In terms of anti-cancer activity, anwuligan could induce apoptosis of human promyeloytic leukemia cells (HL-60) by activating caspase-3 [7], it also exhibits the anti-carcinogenic activity by strongly inhibiting the carcinogenic oral bacteria *S. mutans* [5] and ameliorating the side effects of cancer chemotherapy by regulating the activity of multidrug-resistant protein [8–10]. In terms of hepato-protection activity, anwuligan has been proved to be a drug-metabolizing enzyme (DME) modifie, possessing the ability to inhibit aminopyrin-N-demethylase [11] and ameliorate the cytotoxicity induced by tert-butyl hydroperoxides (t-BHP) in human hepatoma cells (HepG2) [1], anwuligan also has a protective effect on cisplatin-induced hepatotoxicity, which is related to the mitogen activated protein kinase (MAPK) signaling pathway [12]. In terms of anti-diabetic function, anwuligan has an insulin secretagogue action and maintains blood glucose level by inhibiting the intestinal ALPHA-glucoamylase [13], it is also a natural peroxisome proliferator-activated receptor (PPAR)α/γ dual agonist contributed to the complementary and synergistic increases in lipid metabolism and insulin sensitivity [14]. In addition, anwuligan acts on dermalogical protection and has the potential for the treatment of abnormal pigmentation, aging, and related skin diseases [15,16]. In terms of anti-oxidative and anti-inflammatory activities, anwuligan protects mammalian splenocytes from radiation-induced intracellular reactive oxygen species (ROS) production [17]; it also reduces the expression of pro-inflammatory cytokines, TNF (tumor necrosis factor)-α, IL-6, IL-1b, and C-reactive protein [18] and inhibits histamine degranulation [19]. Anwuligan was reported to inhibit the neurotoxicity of glutamate on HT22 cells in the hippocampus of murine and the production of corticotrophin, and have an effect on inflammatory memory loss in chronic LPS rats showing the potential to treat neurodegenerative diseases [20,21]. Toxicity studies have shown that anwuligan has no cytotoxic effect on lymphocytes in vitro [17].

Based on the above results of pharmacological and toxicological studies, it could be inferred that anwuligan is a candidate drug for the treatment of many diseases, which represents a new potential pathway for natural intervention and shows advantages over synthetic chemistry. As far as we know, the study of anwuligan analysis are mainly on HPLC and UV method [22,23], and both are in vitro content determination studies. At present, the pharmacokinetics and tissue distribution of anwuligan have not been reported. It is generally acknowledged that the main purpose of LC-MS bioanalysis is to provide quantitative methods for the candidate drugs, so as to come to accurate and reliable conclusions of pharmacokinetic and tissue distribution studies. Therefore, this study is to establish an accurate and reliable LC-MS/MS method for the determination of anwuligan in rat plasma and tissues, and to be successfully applied in the pharmacokinetic and tissue distribution study, which would provide data support and reference for the future evaluation of preclinical safety of anwuligan as a candidate drug.

2. Results

2.1. Method Validation

2.1.1. Specificity

As shown in Figure 1, the retention time for anwuligan and IS were 4.72 and 3.66 min, respectively. No endogenous interference was observed at the retention time of anwuligan and IS.

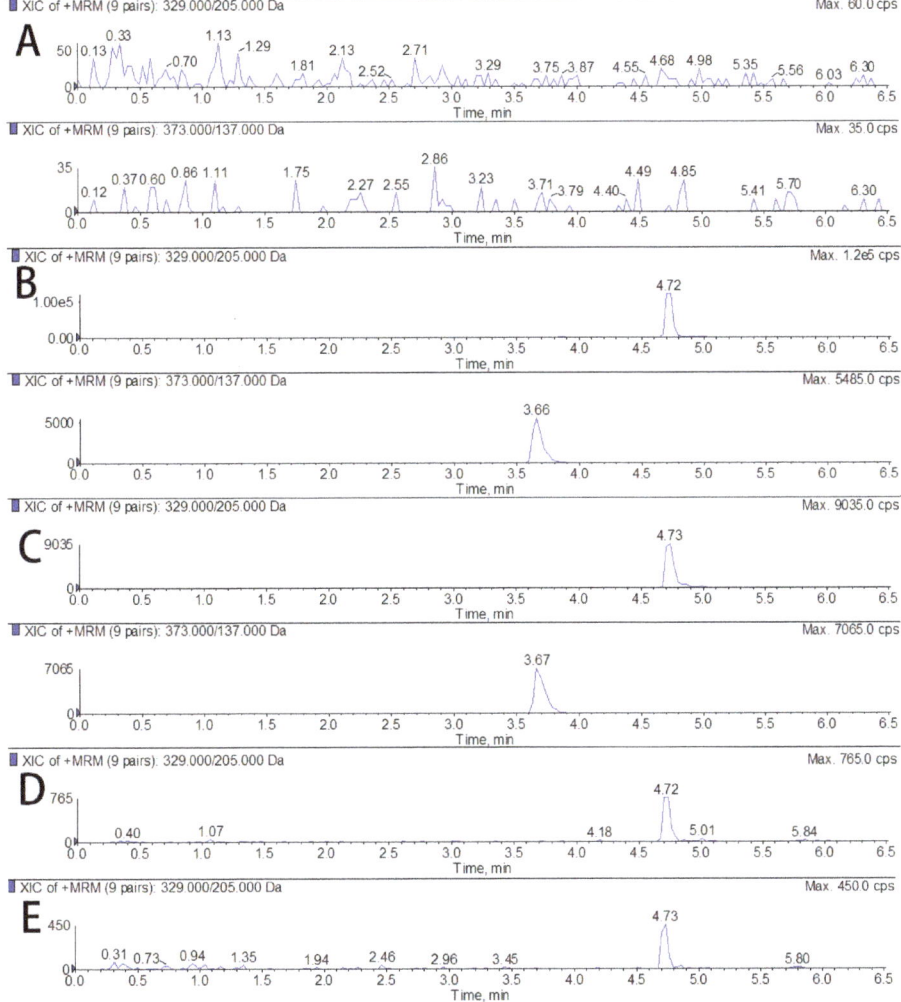

Figure 1. Representative MRM chromatograms of anwuligan: (A) blank plasma; (B) blank plasma spiked with anwuligan and IS; (C) plasma sample collected 1 h after oral administration of anwuligan (60 mg/kg); (D) blank plasma spiked with anwuligan (LLOQ); and (E) blank liver tissue homogenates with anwuligan (LLOQ).

2.1.2. Calibration Curve and LLOQ

The calibration curves were calculated by linear regression method for plasma and tissue samples. As shown in Table 1, the linear concentration range was from 0.5 ng/mL to 2000 ng/mL for plasma and tissue samples, with the coefficient values of more than 0.99.

2.1.3. Precision and Accuracy

Table 2 shows the results of intra-day and inter-day precision and accuracy of anwuligan in rat plasma and tissues. The precision expressed as RSD were all within 13% and the accuracy values expressed as RE ranged from −12.6% to 10.6%, indicating that the method was accurate and reproducible to the quantitative determination of anwuligan in rat plasma and tissues.

Table 1. Calibration curves for anwuligan in biological samples.

Samples	Calibration Curves	Coefficients	SES*	SEI#	Linear Range (ng/mL)	LLOQ (ng/mL)
Plasma	Y = 0.00249 + 0.00157x	0.996–0.997	10⁻⁴	0.22	0.5–2000	0.5
Intestine	Y = 0.00378 + 0.0030x	0.996–0.998	9.6 × 10⁻⁴	0.37	0.5–2000	0.5

Table 1. Calibration curves for anwuligan in biological samples.

Samples	Calibration Curves	Correlation Coefficients (r)	SES *	SEI #	Linear Ranges (ng/mL)	LLOQs (ng/mL)
Plasma	Y = 0.00929 + 0.000157x	0.996–0.997	6.3×10^{-4}	0.22	0.5–2000	0.5
Intestine	Y = 0.00378 + 0.0030x	0.996–0.998	9.6×10^{-4}	0.37	0.5–2000	0.5
Heart	Y = 0.000419 + 0.0491x	0.995–0.997	8.7×10^{-4}	0.19	0.5–2000	0.5
Liver	Y = 0.00587 + 0.000116x	0.997–0.999	4.1×10^{-5}	0.13	0.5–2000	0.5
Spleen	Y = 0.0000987 + 0.0011x	0.994–0.996	7.9×10^{-4}	0.28	0.5–2000	0.5
Lung	Y = 0.00689 + 0.000892x	0.995–0.997	9.0×10^{-4}	0.39	0.5–2000	0.5
Kidney	Y = −0.000326 + 0.000131x	0.996–0.998	7.7×10^{-5}	0.27	0.5–2000	0.5
Stomach	Y = −0.0045 + 0.00088x	0.996–0.997	6.9×10^{-4}	0.34	0.5–2000	0.5
Brain	Y = −0.0068 + 0.00017x	0.996–0.998	8.2×10^{-4}	0.26	0.5–2000	0.5

* SES—Standard error of the slope ($n = 6$). # SEI—Standard error of the intercept ($n = 6$).

Table 2. Precision and accuracy of anwuligan in rat plasma and tissues ($n = 5$).

Samples	QC Conc. (ng/mL)	Intra–Day		Inter–Day	
		Precision (RSD, %)	Accuracy (Mean, %)	Precision (RSD, %)	Accuracy (Mean, %)
Plasma	0.5	11.3	−6.7	7.2	5.3
	5	9.6	−5.3	8.3	6.5
	100	5.3	3.5	4.5	−5.2
	1600	8.4	4.9	3.8	5.7
Heart	0.5	7.5	9.5	6.0	10.6
	5	5.4	−7.0	4.0	−9.2
	100	8.5	7.3	7.0	6.7
	1600	8.7	−3.4	4.1	−4.4
Liver	0.5	4.5	−4.7	9.7	−5.5
	5	8.6	6.4	8.8	9.3
	100	7.1	−5.4	3.9	5.9
	1600	5.3	3.2	6.3	−7.4
Spleen	0.5	8.7	−7.9	9.0	−8.9
	5	4.2	4.0	9.3	−4.7
	100	7.1	−4.4	5.7	−7.1
	1600	6.6	6.9	8.8	3.6
Lung	0.5	6.0	7.5	8.0	6.9
	5	4.9	4.4	5.7	−5.0
	100	4.7	−7.1	3.2	3.8
	1600	6.4	−7.2	6.9	6.6
Kidney	0.5	10.9	−11.5	10.6	−12.6
	5	8.4	−4.3	4.9	−5.1
	100	8.5	5.5	9.0	7.6
	1600	5.4	6.7	5.9	−6.8
Brain	0.5	7.0	−9.3	11.3	−8.7
	5	6.3	−5.1	9.7	−5.5
	100	5.5	−4.4	5.3	3.9
	1600	3.9	3.6	4.2	3.4
Intestine	0.5	6.6	8.6	5.9	8.6
	5	4.6	4.3	3.2	7.2
	100	4.0	6.8	8.1	7.6
	1600	8.3	5.4	4.8	3.1
Stomach	0.5	8.7	6.1	6.9	−8.3
	5	7.8	6.6	6.4	−4.0
	100	5.9	−6.2	4.3	3.9
	1600	7.7	−6.2	7.5	−4.4

2.1.4. Extraction Recovery and Matrix Effect

As presented in Table 3, the extraction recoveries for anwuligan were within 84.0%–114.5%, and the matrix effects for anwuligan were within 84.5% to 96.8%. The results indicated that the extraction efficiency was high and there was little matrix effect for the analyte.

Table 3. Matrix effect and extraction recovery of anwuligan in rat plasma and tissues ($n = 5$).

Samples	QC Conc.(ng/mL)	Matrix Effect		Extract Recovery	
		Mean ± SD (%)	RSD (%)	Mean ± SD (%)	RSD (%)
Plasma	0.5	90.7 ± 2.3	2.6	85.1 ± 2.4	2.9
	5	89.5 ± 7.6	8.6	95.5 ± 5	5.3
	100	96.2 ± 6.3	6.6	109.4 ± 8.5	7.8
	1600	93.1 ± 7.5	8.1	106.4 ± 9.5	8.9
Heart	0.5	90.9 ± 8.8	9.7	86.2 ± 5	5.9
	5	96.6 ± 9.3	9.7	103 ± 2.2	2.1
	100	90.8 ± 4.1	4.6	96.6 ± 9.9	10.3
	1600	84.6 ± 6.3	7.5	84 ± 3.9	4.7
Liver	0.5	93.3 ± 8.3	9.0	109.8 ± 2.4	2.3
	5	89.6 ± 7.2	8.1	106.3 ± 6.4	6.1
	100	85.1 ± 6.0	7.1	114 ± 2.8	2.5
	1600	84.5 ± 9.2	11.0	96.9 ± 9.6	9.9
Spleen	0.5	89.2 ± 6.5	7.3	104.6 ± 4.9	4.7
	5	91.9 ± 4.2	4.6	108.8 ± 3.9	3.6
	100	86.3 ± 7.3	8.5	111.3 ± 2.3	2.1
	1600	96.8 ± 6.5	6.7	90.7 ± 8.8	9.7
Lung	0.5	87.5 ± 5.8	6.7	94.1 ± 3.9	4.2
	5	93.8 ± 2.2	2.4	93.3 ± 8.1	8.8
	100	95.0 ± 9.0	9.5	101.9 ± 9.9	9.7
	1600	88.3 ± 9.8	11.1	102.2 ± 2.8	2.7
Kidney	0.5	86.7 ± 3.9	4.6	95.9 ± 2.4	2.6
	5	87.2 ± 8.7	10.1	114.5 ± 8.4	7.4
	100	87.4 ± 9.8	11.2	105.2 ± 5	4.8
	1600	90.5 ± 4.3	4.8	84.5 ± 4.5	5.4
Brain	0.5	88.5 ± 3.7	4.2	102.5 ± 5.8	5.7
	5	95.8 ± 4.5	4.7	84.8 ± 4.9	5.8
	100	95.1 ± 7.0	7.4	103.6 ± 5.4	5.3
	1600	85.5 ± 5.7	6.7	106.9 ± 6.4	6.0
Intestine	0.5	85.8 ± 3.9	4.6	93 ± 4.7	5.1
	5	86.0 ± 2.8	3.4	104.9 ± 5.4	5.2
	100	84.6 ± 8.7	10.3	102 ± 9.8	9.6
	1600	92.8 ± 8.1	8.8	92.8 ± 7.1	7.6
Stomach	0.5	91.8 ± 6.9	7.5	114.5 ± 8.1	7.1
	5	88.0 ± 2.4	2.8	96.7 ± 2.8	2.9
	100	87.8 ± 9.0	10.3	93.8 ± 9	9.6
	1600	91.7 ± 3.7	4.0	102 ± 3.4	3.4

2.1.5. Stability

Table 4 shows the results of stability of anwuligan under the four given storage conditions, which indicated there was no stability issue occurred.

2.1.6. Dilution Intergrity

Dilution integrity was assessed by six replicate samples, with the plasma concentration of 20.0 μg/mL and tissue homogenate concentration of 12.5 μg/mL (50.0 μg/g tissue) for anwuligan, were diluted 20-fold with blank rat plasma and blank tissue homogenate respectively, which was within the range of standard curve. The precision (CV) was less than 15% and the accuracy was within 85%–115% for the analyte.

Table 4. Stability results for anwuligan in rat plasma and tissues homogenates of rats ($n = 5$).

Samples	QC Conc.(ng/mL)	Short-Term (at Room Temperature for 4 h)	Autosampler 4 °C for 24 h	Three Freeze-Thaw Cycles	Storage at −80 °C for 30 d
Plasma	5	110.4 ± 3.2	114.4 ± 10.7	110.0 ± 4.7	106.7 ± 3.3
	1600	94.2 ± 10.4	85.2 ± 5.9	86.0 ± 4.3	99.0 ± 5.6
Heart	5	95.0 ± 10.4	90.3 ± 10.6	107.1 ± 4.5	85.2 ± 4.5
	1600	94.8 ± 5.2	102.6 ± 7.8	111.8 ± 7.3	96.0 ± 3.3
Liver	5	97.3 ± 4.8	91.6 ± 8.0	105.6 ± 9.1	103.2 ± 7.7
	1600	87.1 ± 8.9	87.0 ± 7.5	106.7 ± 9.5	86.4 ± 6.3
Spleen	5	98.5 ± 10.2	100.9 ± 5.3	98.6 ± 10.5	110.0 ± 8.2
	1600	105.6 ± 5.3	87.6 ± 5.1	102.6 ± 7.7	114.9 ± 6.2
Lung	5	112.4 ± 3.7	97.2 ± 3.5	88.6 ± 7.6	111.1 ± 4.8
	1600	87.2 ± 9.7	93.6 ± 4.0	112.9 ± 9.7	92.8 ± 7.5
Kidney	5	95.7 ± 9.9	95.2 ± 6.3	88.7 ± 6.4	86.4 ± 8.4
	1600	99.8 ± 9.1	97.1 ± 6.2	104.3 ± 9.3	112.2 ± 6.3
Brain	5	93.2 ± 7.4	104.5 ± 4.2	90.0 ± 5.0	112.8 ± 7.7
	1600	104.1 ± 4.0	106.2 ± 3.3	106.9 ± 8.7	112.5 ± 7.1
Intestine	5	103.8 ± 4.9	95.5 ± 6.6	87.1 ± 4.1	95.1 ± 9.6
	1600	111.6 ± 10.1	110.3 ± 7.1	108.3 ± 7.4	113.9 ± 4.9
Stomach	5	90.2 ± 8.0	99.9 ± 3.2	88.8 ± 4.3	105.9 ± 7.5
	1600	104.0 ± 10.1	109.4 ± 9.8	107.2 ± 9.4	89.3 ± 8.7

2.2. Dose Selection

The intravenous administration dose was generally determined based on 1/20–1/50 of the LD50 value or 1/2–1/3 of the maximum tolerated dose of the reference MTD [24–26]. The LD50 obtained by our previous pharmacological experiment was 573 mg/kg and therefore the selected i.v. dose was 5 mg/kg. Meanwhile, it is reported that the structural analog of anwuligan, was given to rat i.v. at a dose of 20 mg/kg for pharmacokinetic study [27,28], was given to rat i.g. at a dose of 40 mg/kg for pharmacokinetic study [28]. The intravenous dosage of another piece of literature is 20 mg/kg for pharmacological activity study [29]. Above all, the dosage of drug intravenous and intragastric administration in our pharmacokinetics experiment is determined to be 15 mg/kg and 60 mg/kg, respectively. In addition, the bioavailability of different routes of administration is different. Generally, the bioavailability of intravenous administration is 100%, oral administration is about 25%~30%, and the ratio is 1/4. The calculated bioavailability was 16%, which was close to 25%, so the ratio of the two routes was more reasonable.

2.3. Pharmacokinetic Study

The validated LC-MS/MS method has been successfully applied to the pharmacokinetics and tissue distribution study of anwuligan after the intravenous or intragastric administration. The mean plasma concentration–time profile of anwuligan after administration was shown in Figure 2. The pharmacokinetic parameters based on the non-compartmental method were summarized in Table 5.

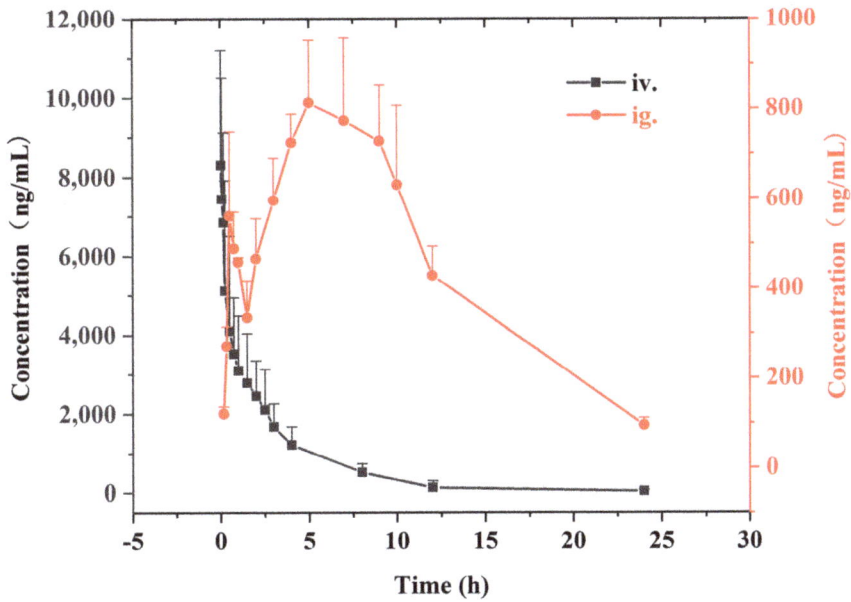

Figure 2. The plasma concentration-time course in rats by given i.v. dose of 15 mg/kg of anwuligan (n = 6) and given i.g. dose of 60 mg/kg of anwuligan (n = 6).

Table 5. Non-compartmental pharmacokinetic parameters of anwuligan.

Pharmacokinetic Parameters	Intravenous	Intragastric
C_{max} (ng/mL)	8310 ± 910	810 ± 190
$t_{1/2}$ (h)	3.20 ± 1.09	5.17 ± 1.82
AUC_{0-t} (ng/mL·h)	16,700 ± 7700	10,700 ± 2800
$AUC_{0-\infty}$ (ng/mL·h)	16,800 ± 8500	11,400 ± 3600
$MRT_{0-\infty}$ (h)	3.71 ± 0.96	10.02 ± 2.04
CL (L/h/kg)	0.89 ± 0.32	5.26 ± 1.39
V_d (L/kg)	4.1 ± 1.1	39.3 ± 12.6
F (bioavailability, %)	-	16.2

2.3.1. Intravenous Administration PKs

Upon intravenous administration of 15 mg/kg anwuligan, the C_{max} was 8310 ± 910 ng/mL, $t_{1/2}$ was 3.20 ± 1.09 h, indicating that anwuligan was rapidly eliminated in plasma. The pharmacokinetic results have demonstrated the area under the curve up to the last sampling time (AUC_{0-t}) and the infinite time ($AUC_{0-\infty}$) of 16,700 ± 7700 and 16,800 ± 8500 ng/mL·h, respectively. The total mean residence time up to the last sampling time (MRT_{0-t}) was found to be 3.71 ± 0.96 h. Anwuligan had an apparent V_d of 4.1 ± 1.1 L/kg which showed that the drug had no accumulation in tissues.

2.3.2. Intragastric Administration PKs

Upon intragastric administration of 60 mg/kg anwuligan, the C_{max} was 810 ± 190 ng/mL. The drug was rapidly absorbed and the $t_{1/2}$ was 5.17 ± 1.82 h, indicating that the drug was relatively slowly eliminated. The pharmacokinetic results have demonstrated the area under the curve up to time (AUC_{0-t}) and the infinite time ($AUC_{0-\infty}$) of 10,700 ± 2800 and 11,400 ± 3600 ng/mL·h, respectively. Anwuligan had an apparent V_d of 39.3 ± 12.6 L/kg and CL of 5.26 ± 1.39 L/h/kg. The total mean residence time up to time (MRT_{0-t}) was found to be 10.02 ± 2.04 h. The absolute bioavailability is 16.2%, which represented the existing of the obvious first-pass effect. As shown in the Figure 2, residence time up to time (MRT$_{0-t}$) was found to be 10.02 ± 2.04 h. The absolute bioavailability is 16.2%, which represented the existing of the obvious first-pass effect. As shown in the Figure 2, the drug concentration increased again after 2 h in the oral administration mode, presumably an enterohepatic circulation.

2.4. Tissue Distribution

The change trend of anwuligan concentration in different tissues at different time points at 0.5, 1, 2, and 4 h after administration was observed. As shown in Figure 3, anwuligan was distributed widely and rapidly in different tissues.

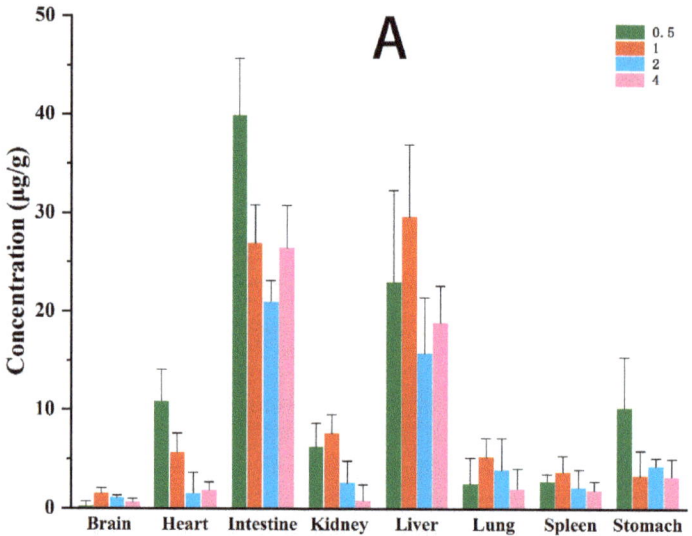

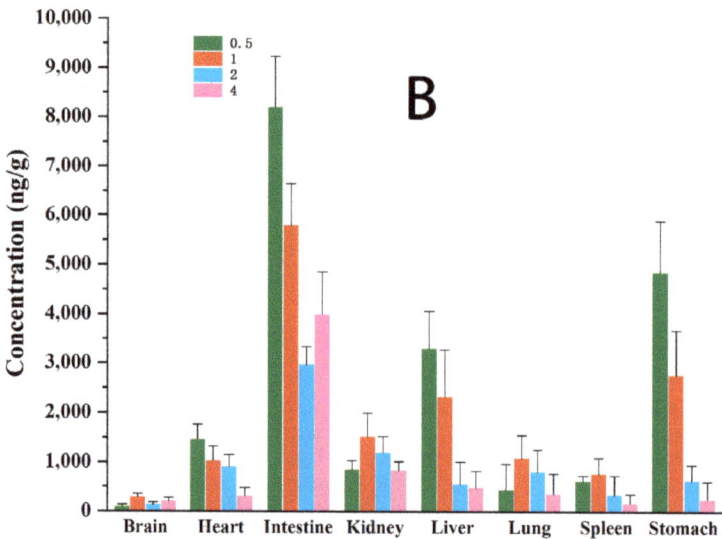

Figure 3. Mean concentration of anwuligan in various tissues at 0.5, 1, 2 and 4 h (**A**) given i.v. dose of 15 mg/kg of anwuligan ($n = 4$) and (**B**) given i.g. dose of 60 mg/kg of anwuligan ($n = 4$).

2.4.1. Intravenous Administration Tissue Distribution

In brain, lung, and spleen, the maximum concentration appeared at 1 h, while in kidney, heart, liver, intestine, and stomach tissues, the maximum concentration appeared at 0.5 h. Among all the tissues, the liver and intestine showed the highest average concentration, which was just a proof for the enterohepatic circulation thought in pharmacokinetic study.

2.4.2. Intragastric Administration Tissue Distribution

The maximum concentration of stomach, kidney and intestine tissue appeared at 1 h, the maximum concentration of heart and lung tissue appeared at 2 h, while the maximum concentration of brain and liver tissue appeared at 4 h. Among all the tissues, the average concentration of liver, stomach and intestine was the highest, which proved that there was an enterohepatic circulation. The high concentration of drugs in the stomach and intestine was due to the oral administration mode. Therefore, anwuligan maintained a high concentration in the liver, suggesting anwuligan might undergo a relative slow elimination and long accumulation in the liver. It was reported anwuligan could inhibit cyp450 activity [10] leading to weaken the activity of drug enzymes and slow down the metabolism of their own or other drugs, which could well explain the slow elimination and long accumulation of anwuligan in the liver. Moreover it was reported that the existing of methylenedioxyphenyl moiety in anwuligan indicated it might have the ability in inhibiting aminopyrine-N-demethylase [11], suggesting a plausible hepatoprotective mechanism. Therefore, liver was very likely to be the target organ of curative effect of anwuligan, which requires systematic research in the future. In contrast, anwuligan was less likely to cause drug accumulation and immune side effects in heart, kidney, brain, lung, and spleen due to the rapid elimination and less drug accumulation. At the same time, anwuligan could be detected in the brain, which indicated that it was able to pass through the blood-brain barrier.

3. Materials and Methods

3.1. Reagents and Materials

Anwuligan (purity over 99%) and arctigenin (IS, purity over 99%) were purchased from Chengdu Biopurity Phytochemicals Ltd. (Chengdu, Sichuan, China). Acetonitrile and methanol were MS-grade reagents, purchased from Merck KGaA Company (Darmstadt, Germany) and formic acid was HPLC-grade, purchased from Dikma Company (Lake Forest CA, USA). Deionized water was purified via a Millipore Milli-Q system (Millipore, Bedford, MA, USA). The other chemical reagents were of analytical grade. The chemical structures of anwuligan and IS are shown in Figure 4.

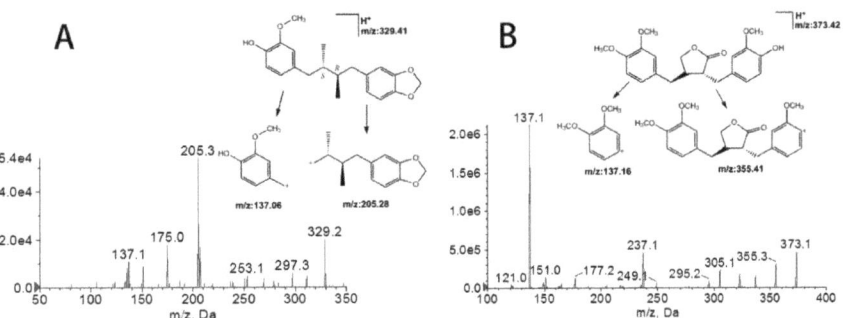

Figure 4. Product ion spectra of [M+H]+ for anwuligan (A) and arctigenin (IS, B).

3.2. Animals

Adult male Sprague–Dawley rats (SD, 200 ± 20 g) were obtained from Experimental Animal Research Center, China Medical University (Shenyang, China) and housed in a temperature and

3.4. Preparation of Calibration Standards and Quality Control (QC) Samples

The stock solution of anwuligan (100 μg/mL) was prepared in acetonitrile, and serially diluted to the working solutions with acetonitrile. The IS working solution with a concentration of 250 ng/mL was also prepared in acetonitrile. The plasma calibration standards were prepared by spiking 50 μL of the working solutions of anwuligan and 50 μL of IS (250 ng/mL) into 100 μL rat blank plasma (or blank tissue homogenate). The final concentrations of anwuligan in rat plasma and tissue homogenates were both ranged from 0.5 to 2000 ng/mL (0.5, 1, 4, 10, 40, 160, 400, 800, and 2000 humidity controlled room at 24 ± 2 °C and 50 ± 10%. Water and food were available ad libitum. The rats were fed for two weeks and fasted for 12 h before treatment. The whole experimental protocol was approved by the Institutional Animal Care and Use Committee at China Medical University (CMU2019194).

3.3. LC-MS/MS Conditions

The biological samples were analyzed using an Agilent series 1290 UHPLC system (Agilent Technologies, Santa Clara, CA, USA) equipped with an AB 3500 triple quadrupole mass spectrometer (AB Sciex, Ontario, ON, Canada) and an electrospray ionization (ESI) source. The separation was carried out on an ACQUITY UPLC BEH C_{18} Column (100 mm × 2.1 mm, 1.7 μm) from Agilent Technologies (Santa Clara, CA, USA) at 30 °C. A gradient program was employed with a mobile phase consisting of 0.1% formic acid aqueous solution (A) and acetonitrile (B) at the flow rate of 0.3 mL/min: 0.0–1.0 min (80%A), 1.0–1.5 min (80%–5%A), 1.5–5.0 min (5%–1%A), 5.0–6.0 min (1%A), 6.0–6.5 min (80%A). The total run time was 6.5 min. The injection volume was 10 μL.

The mass spectrometer was operated in a positive ionization mode, using multiple reaction monitoring (MRM) with following operation parameters: Ionspray Voltage, 5500 V; the turbo spray temperature, 500 °C; ion source gas 1, 19; ion source gas 2, 19; curtain gas, 10. Collision cell exit potential (CXP) and entrance potential (EP) was set at 7.0 V and 10.0 V, respectively. Based on the mass scan and product ion scan (Figure 4), the transitions for quantification were m/z 329.0→205.0 for anwuligan and m/z 373.0→137.0 for IS, respectively. Moreover, the qualifier ions for anwuligan and IS were set at m/z 137.0 and 355.0, respectively. The declustering potential (DP) for anwuligan and IS were 80 V and 80 V; the collision energy (CE) were 18 eV and 15 eV, respectively. The 1.6.3 version Analyst software package (AB Sciex, Ontario, ON, Canada) was used for data acquisition and instrument control.

3.4. Preparation of Calibration Standards and Quality Control (QC) Samples

The stock solution of anwuligan (100 μg/mL) was prepared in acetonitrile, and serially diluted to the working solutions with acetonitrile. The IS working solution with a concentration of 250 ng/mL was also prepared in acetonitrile. The plasma calibration standards were prepared by spiking 50 μL of the working solutions of anwuligan and 50 μL of IS (250 ng/mL) into 100 μL rat blank plasma (or blank tissue homogenate). The final concentrations of anwuligan in rat plasma and tissue homogenates were both ranged from 0.5 to 2000 ng/mL (0.5, 1, 4, 10, 40, 160, 400, 800, and 2000 ng/mL). Quality control (QC) samples at 0.5, 5, 100, and 1600 ng/mL were also prepared in the same manner.

3.5. Sample Preparation

3.5.1. Plasma Samples

In this study, a liquid-liquid extraction method was applied to prepare the PKs and tissue distribution samples. An aliquot 100 μL of plasma, 50 μL of IS solution (250 ng/mL) and 1 mL of extraction agent ethyl acetate was added into a 1.5 mL Eppendorf tube. The mixture was vortexed for 1 min. After the centrifugation at 12,000× g for 7 min at 4 °C, a 900 μL aliquot of the upper organic layer was carefully transferred and evaporated to dryness at 40 °C under nitrogen. The residuals were reconstituted in 100 μL of initial mobile phase (acetonitrile-0.1% fomic acid water, 20:80, v/v) by vortex mixing for 30 s. A volume of 10 μL of the supernatant was injected into the LC-MS/MS system for analysis after centrifuging at 12,000× g for 10 min.

3.5.2. Tissue Samples

The rats were sacrificed on the ice. The rat tissues were quickly collected, including the heart, the liver, the spleen, the lung, the kidney, the brain, the stomach, and the intestine, and washed with ice-cold 0.9% NaCl to remove the surface blood. After suction with filter paper, a certain amount of tissue sample was weighed on ice and homogenized with physiological saline at a ratio of 1:4, *w/v*. If the tissue sample was <0.5 g, all tissue was taken; if the tissue sample was 0.5–1.0 g, 0.5 g of tissue sample was taken; if the tissue samples was >1.0 g, 1.0 g of tissue sample was taken. Then, a 100 µL (equivalent to 25 mg) of tissue homogenate was taken and processed in the same manner as shown in "Section 3.5.1".

3.6. Method Validation

The method validation was based on the guidelines set by US Food and Drug Administration (FDA) guidelines [30].

3.6.1. Specificity

The specificity was assessed by comparing the chromatograms of six different lots of rat blank plasma samples, blank plasma (tissue) spiked with anwuligan at LLOQ and IS, and the plasma samples after the administration of anwuligan. There should be no interference at the retention times of the analytes and the IS for blank samples and the retention times should be consistent for spiked samples and actual samples.

3.6.2. Calibration Curve and LLOQ

Calibration curves were constructed using a weighted ($1/x^2$) least square linear regression by plotting the peak area ratios of analyte to IS versus the nominal plasma concentrations of anwuligan over the range of 0.5–2000 ng/mL.

The low limit of quantification (LLOQ), defined as the lowest concentration of the calibration curve, should be quantitatively determined with the precision (expressed as RSD, %) within 20% and accuracy (expressed as RE, %) within ± 20%, and the S/N of which should be over 10.

3.6.3. Precision and Accuracy

The intra-day and inter-day precision and accuracy were evaluated by the determination of QC samples at three concentration levels and LLOQ in six replicates on one day and on three consecutive days, respectively. The precision and accuracy were expressed as RSD (%) and RE (%), and expected to be within ± 15% for QC samples.

3.6.4. Extraction Recovery and Matrix Effect

The recovery of the analyte at three QC levels ($n = 5$) was determined by the peak area ratios of extracted analytes to those of the post-extraction samples containing equivalent amounts of targets. The matrix effect was evaluated by comparing the peak areas of the targets in the spike-after-extraction samples with the mean peak areas of it dissolved with the mobile phase at high, medium, and low levels, respectively.

3.6.5. Stability

The stability of anwuligan was evaluated by analyzing five replicates of the samples at three QC levels, including storage in the auto-sampler (4 °C) for 24 h, 4 h exposure at room temperature, three freeze thaw cycles, frozen at −80 °C for 30 days.

3.6.6. Dilution Integrity

At dilution factor 20-fold, the accuracy of the diluted samples for plasma samples (20 ug/mL diluted to 1000 ng/mL) and tissue samples (12.5 ug/mL diluted to 625 ng/mL) was 5.14%–6.21%, and the precision was 5.53%–6.77%, respectively. These results suggested that the study samples can be diluted and maintain adequate accuracy and precision values.

3.7. PKs and Tissue Distribution Study Protocols

For the pharmacokinetic study, the rats were fasted for 12 h with free access to water prior to administration. Anwuligan was dissolved in 0.5% DMSO saline (v/v) to give the intravenous administration solution at the dose of 15 mg/kg and dissolved in 0.5% DMSO saline (v/v) to give the oral administration solution the dose of 60 mg/kg. For intravenous administration, the approach is using a syringe to deliver the drug into the tail vein; for intragastric administration, the approach is to insert a gastric lavage needle into the mouth of one side of the rat and the drug is slowly injected into the stomach for administration.

A blood sample (0.2 mL) was obtained from the orbital vein and transferred to heparinized tubes at 0.033, 0.083, 0.167, 0.25, 0.50, 0.75, 1, 1.5, 2, 2.5, 3, 4, 8, 10 h after intravenous or oral administration. Six rats are used for each group. The blood samples were immediately centrifuged (12,000× g, 10 min) and stored at −80 °C before analysis.

For the tissue distribution study, 16 SD rats received an oral administration at the dose of 60 mg/kg and another 16 SD rats received an intravenous administration at the dose of 15 mg/kg. After that, the tissue samples of heart, liver, spleen, lung, kidney, brain, stomach, and intestine were collected and weighed at 0.5, 1, 2, and 4 h, respectively. For the preparation method of plasma and tissue homogenate samples (see Section 3.5).

3.8. Data Analysis

The pharmacokinetic parameters were calculated by a non-compartmental model using the DAS 3.2.8 pharmacokinetic program [31].

4. Conclusions

In this study, a rapid, simple, and sensitive UPLC-MS/MS method was developed and validated, following by successfully applied to the pharmacokinetic and tissue distribution investigation of anwuligan after the intravenous or intragastric administration. The method has been validated for the first time in this paper and proved to be able to detect a low concentration of 0.5 ng/mL for anwuligan with liquid-liquid extraction method. The elimination half-life of anwuligan of i.g. administration is relatively shorter than that of i.v. administration. The absolute bioavailability is 16.2%, which represented the existing of the obvious first-pass effect. An enterohepatic circulation was found in the i.g. mode. Anwuligan could be distributed rapidly and widely in different tissues and maintained a high concentration in the liver, suggesting anwuligan might undergo a relative slow elimination and long accumulation in the liver. The developed and validated LC-MS/MS method and the pharmacokinetic study of anwuligan in the present paper would provide data support and reference for the future evaluation of preclinical safety of anwuligan as a candidate drug.

Supplementary Materials: The following are available online, Table S1: Pharmacology of anwuligan.

Author Contributions: All authors contributed to the article as follows: X.-S.F. conceived and designed the experiments; Y.S. wrote the paper; Y.Z. performed the experiments; X.-Y.D., D.-W.C. analyzed the data; X.Q., Y.B., and K.-F.W. performed the chromatographic analysis of plasma and biosamples. All authors read and approved the final manuscript. All authors have read and agreed to the published version of the manuscript.

Funding: This work was supported by Liaoning planning Program of Philosophy and Social Science (no. L17BGL034); Key Program of the Natural Science Foundation of Liaoning Province of China (no. 20170541027); Scientific Research Project of Department of Education of Liaoning Province (ZF2019036); and the Major Subject in Education Research founded by Chinese Medical Association Medical Education Branch and the China Association of Higher Education Medical Education Professional Committee (2018A-N19012).

Conflicts of Interest: The authors declare no conflict of interest.

References

1. Sohn, J.H.; Han, K.L.; Choo, J.H.; Hwang, J.K. Macelignan protects HepG2 cells against tert-butylhydroperoxide-induced oxidative damage. *BioFactors* **2007**, *29*, 1–10. [CrossRef] [PubMed]
2. Woo, S.W.; Shin, K.H.; Wagner, H.; Lotter, H. The structure of macelignan from Myristica fragrans. *Phytochemistry* **1987**, *26*, 1542–1543. [CrossRef]
3. Orabi, Y.K.; Mossa, J.S.; El-Feraly, F.S. Isolation and characterization of two antimicrobial agents from mace (Myristica fragrans). *J. Nat. Prod.* **1991**, *54*, 856–859. [CrossRef] [PubMed]
4. Rukayadi, Y.; Lee, K.; Han, S.; Kim, S.; Hwang, J.K. Antibacterial and Sporicidal Activity of Macelignan Isolated from Nutmeg (Myristica fragrans Houtt.) against Bacillus cereus. *Food Sci. Biotechnol.* **2009**, *18*, 1301–1304.
5. Chung, J.Y.; Choo, J.H.; Lee, M.H.; Hwang, J.K. Anticariogenic activity of macelignan isolated from Myristica fragrans (nutmeg) against Streptococcus mutans. *Phytomedicine* **2006**, *13*, 261–266. [CrossRef] [PubMed]
6. Yan, T.; Rukayadi, Y.; Kim, K.H.; Hwang, J.K. In Vitro Anti-biofilm Activity of Macelignan isolated from Myristica fragrans Houtt. against Oral Primary Colonizer Bacteria. *Phytother. Res.* **2008**, *22*, 308–312.
7. Park, B.Y.; Min, B.S.; Kwon, O.K.; Oh, S.R.; Ahn, K.S.; Kim, T.J.; Kim, D.Y.; Bae, K.; Lee, H.K. Increase of caspase-3 activity by lignans from Machilus thunbergii in HL-60 cells. *Biol. Pharm. Bull.* **2004**, *27*, 1305–1307. [CrossRef]
8. Qiang, F.; Lee, B.; Ha, I.; Kang, K.W.; Woo, E.; Han, H. Effect of maceligan on the systemic exposure of paclitaxel: In vitro and in vivo evaluation. *Eur. J. Pharm. Sci.* **2010**, *41*, 226–231. [CrossRef]
9. Zhang, C.; Qi, X.; Shi, Y.; Sun, Y.; Li, S.; Gao, X.; Yu, H. Estimation of Trace Elements in Mace (Myristica fragrans Houtt) and Their Effect on Uterine Cervix Cancer Induced by Methylcholanthrene. *Biol. Trace Elem. Res.* **2012**, *149*, 431–434. [CrossRef]
10. Im, Y.B.; Ha, I.; Kang, K.W.; Lee, M.; Han, H. Macelignan: A New Modulator of P-Glycoprotein in Multidrug-Resistant Cancer Cells. *Nutr. Cancer* **2009**, *61*, 538–543. [CrossRef]
11. Shin, S.H.; Woo, W.S.; Lee, J.Y.; Han, Y.B. Effects of Lignans on Hepatic Drug-Metabolizing Enzymes. *Arch. Pharmacal. Res.* **1990**, *13*, 265–268. [CrossRef]
12. Sohn, J.H.; Han, K.L.; Kim, J.H.; Rukayadi, Y.; Hwang, J.K. Protective Effects of macelignan on cisplatin-induced hepatotoxicity is associated with JNK activation. *Biol. Pharm. Bull.* **2008**, *31*, 273–277. [CrossRef] [PubMed]
13. Patil, S.B.; Ghadyale, V.A.; Taklikar, S.S.; Kulkarni, C.R.; Arvindekar, A.U. Insulin Secretagogue, Alpha-glucosidase and Antioxidant Activity of Some Selected Spices in Streptozotocin-induced Diabetic Rats. *Plant Foods Hum. Nutr.* **2011**, *66*, 85–90. [CrossRef] [PubMed]
14. Staels, B.; Fruchart, J.C. Original Article Therapeutic Roles of Peroxisome Proliferator- Activated Receptor Agonists. *Diabetes* **2005**, *54*, 2460–2470. [CrossRef] [PubMed]
15. Cho, Y.; Kim, K.H.; Shim, J.S.; Hwang, J.K. Inhibitory effects of macelignan isolated from Myristica fragrans HOUTT. On melanin biosynthesis. *Biol. Pharm. Bull.* **2008**, *31*, 986–989. [CrossRef] [PubMed]
16. Choi, E.J.; Kang, Y.G.; Kim, J.; Hwang, J.K. Macelignan inhibits melanosome transfer mediated by protease-activated receptor-2 in keratinocytes. *Biol. Pharm. Bull.* **2011**, *34*, 748–754. [CrossRef]
17. Checker, R.; Chatterjee, S.; Sharma, D.; Gupta, S.; Variyar, P.; Sharma, A.; Poduval, T.B. Immunomodulatory and radioprotective effects of lignans derived from fresh nutmeg mace (Myristica fragrans) in mammalian splenocytes. *Int. Immunopharmacol.* **2008**, *8*, 661–669. [CrossRef]
18. Han, K.L.; Choi, J.S.; Lee, J.Y.; Song, J.; Joe, M.K.; Jung, M.H.; Hwang, J. Therapeutic Potential of Peroxisome Proliferators-Activated Receptor-α/γ Dual Agonist With Alleviation of Endoplasmic Reticulum Stress for the Treatment of Diabetes. *Diabetes* **2008**, *57*, 737–745. [CrossRef]
19. Han, Y.S.; Kim, M.; Hwang, J. Macelignan Inhibits Histamine Release and Inflammatory Mediator Production in Activated Rat Basophilic Leukemia Mast Cells. *Inflammation* **2012**, *35*, 1723–1731. [CrossRef]
20. Jin, D.; Lim, C.S.; Hwang, J.K.; Ha, I.; Han, J. Anti-oxidant and anti-inflammatory activities of macelignan in murine hippocampal cell line and primary culture of rat microglial cells. *Biochem. Biophys. Res. Commun.* **2005**, *331*, 1264–1269. [CrossRef]
21. Cui, C.; Jin, D.; Hwang, Y.K.; Lee, I.; Hwang, J.K.; Ha, I.; Han, J. Macelignan attenuates LPS-induced inflammation and reduces LPS-induced spatial learning impairments in rats. *Neurosci. Lett.* **2008**, *448*, 110–114. [CrossRef] [PubMed]

22. Jiang, P.; Lu, Y.; Chen, D. Authentication of Schisandra chinensis and Schisandra sphenanthera in Chinese patent medicines. *J. Pharm. Biomed. Anal.* **2016**, *131*, 263–271. [CrossRef] [PubMed]
23. Wei, H.; Sun, L.; Tai, Z.; Gao, S.; Xu, W.; Chen, W. A simple and sensitive HPLC method for the simultaneous determination of eight bioactive components and fingerprint analysis of Schisandra sphenanthera. *Anal. Chim. Acta* **2010**, *662*, 97–104. [CrossRef]
24. Chen, Q. *Research Methods in Pharmacology of Chinese Materia Medica*, 3rd ed.; People's Medical Publishing House: Beijing, China, 2011.
25. Sharma, D.R.; Sunkaria, A.; Bal, A.; Bhutia, Y.D.; Vijayaraghavan, R.; Flora, S.J.; Gill, K.D. Neurobehavioral impairments, generation of oxidative stress and release of pro-apoptotic factors after chronic exposure to sulphur mustard in mouse brain. *Toxicol. Appl. Pharm.* **2009**, *15*, 208–218. [CrossRef]
26. Tao, X.; Cush, J.J.; Garret, M.; Lipsky, P.E. A phase I study of ethyl acetate extract of the chinese antirheumatic herb Tripterygium wilfordii hook F in rheumatoid arthritis. *J. Rheumatol.* **2005**, *10*, 2160–2167.
27. Mukker, J.K.; Kotlyarova, V.; Singh, R.S.P.; Alcorn, J. HPLC method with fluorescence detection for the quantitative determination of flaxseed lignans. *J. Chromatogr. B* **2010**, *878*, 3076–3082. [CrossRef]
28. Mukker, J.K.; Singh, R.S.P.; Muir, A.D.; Krol, E.S.; Alcorn, J. Comparative pharmacokinetics of purified flaxseed and associated mammalian lignans in male Wistar rats. *Br. J. Nutr.* **2015**, *113*, 749–757. [CrossRef] [PubMed]
29. Li, F.; Yang, X. Quantification of myrislignan in rat plasma by solid-phase extraction and reversed-phase high-performance liquid chromatography. *Biomed. Chromatogr.* **2008**, *22*, 601–605. [CrossRef]
30. Bioanalytical Method Validation Guidance for Industry 2018. Available online: https://www.fda.gov/files/drugs/published/Bioanalytical-Method-Validation-Guidance-for-Industry.pdf (accessed on 2 November 2019).
31. Drug Analysis System [CP/DK]. Available online: http://www.drugchina.net/# (accessed on 2 November 2019).

Sample Availability: Samples of the anwuligan and arctigenin are available from the authors.

© 2019 by the authors. Licensee MDPI, Basel, Switzerland. This article is an open access article distributed under the terms and conditions of the Creative Commons Attribution (CC BY) license (http://creativecommons.org/licenses/by/4.0/).

Article

Interaction Effects between Doxorubicin and Hernandezine on the Pharmacokinetics by Liquid Chromatography Coupled with Mass Spectrometry

Yang Song [1,†], Yuan Zhang [2,†], Wei-Peng Zhang [1], Bao-Zhen Zhang [1], Ke-Fei Wang [1] and Xue-Song Feng [1,*]

1. School of Pharmacy, China Medical University, Shenyang 110122, China; songyanglhyb1998@163.com (Y.S.); wpzhang@cmu.edu.cn (W.-P.Z.); piaotaihengmartin@163.com (B.-Z.Z.); wl5712367805@sina.com (K.-F.W.)
2. Department of Pharmacy, National Cancer Center/National Clinical Research Center for Cancer, Chinese Academy of Medical Sciences and Peking Union Medical College, Beijing 100021, China; zhangyuan@cicams.ac.cn
* Correspondence: Correspondence: xsfeng@cmu.edu.cn; Tel.: +86-189-0091-1848; Fax: +86-024-3193-9448
† These authors contributed equally to this work.

Received: 11 September 2019; Accepted: 4 October 2019; Published: 8 October 2019

Abstract: Doxorubicin (DOX) is an effective anti-tumor drug widely used in clinics. Hernandezine (HER), isolated from a Chinese medicinal herb, has a selective inhibitory effect on DOX multidrug resistance, making DOX more effective in treating cancer. The aim of this study was to investigate the effect of the interaction of HER and DOX on pharmacokinetics. Male Sparague–Dawley rats were randomly divided into three groups: a single DOX group, a single HER group, and a combination group. Plasma concentrations of DOX and HER were determined by the LC-MS/MS method at specified time points after administration, and the main pharmacokinetic parameters were estimated. The results showed that there were significant differences in the C_{max} and $AUC_{0-\infty}$ of DOX in the single drug group and combined drug group, indicating that HER could improve the absorption of DOX. However, DOX in combination, in turn, reduced the free drug concentration of HER, possibly because DOX enhanced the HER drug–protein binding effect. The results could be used as clinical guidance for DOX and HER to avoid adverse reactions.

Keywords: doxorubicin; hernandezine; LC-MS/MS; pharmacokinetic study; drug–drug interaction

1. Introduction

Doxorubicin (DOX), as an anti-tumor drug widely used in clinical treatment, has an obvious curative effect on various tumors, including leukemia, malignant lymphoma, and various solid tumors [1–3]. However, its long-term or high-dose clinical use often leads to irreversible congestive heart failure, which limits its application [4,5]. In previous studies, the cardiotoxicity may be related to the formation of free radicals, lipid peroxidation, Ca^{2+} overloading, and the activation of apoptotic factors [6,7]. In order to find an effective way to solve this problem, many research studies have been carried out, but the results were not satisfactory [8]. The multidrug resistance (MDR) of DOX is a major obstacle to its application in tumor chemotherapy. Although various mechanisms are known to be involved in MDR phenotypes, the overexpression of some members of the ATP-binding cassette (ABC) protein family is considered to be a major contributor to MDR development in tumor cells [9].

Hernandezine (HER), a dibenzyl isoquinoline alkaloid isolated from traditional Chinese medicine, has long been used in the treatment of hypertension [10,11]. HER has been proved to be able to prevent hair cell aminoglycoside-induced injury [12], inhibit protein kinase C signal events in human peripheral blood T cells [13] and neuronal nicotinic acetylcholine receptors (nAChRs), [14], and block non-voltage-operated Ca^{2+} entry activated by intracellular Ca^{2+} store depletion induced by thapsigargin in rat glioma C6 cells [15] and in human leukemic HL-60 cells [16]. In addition, HER was found to be an effective MDR modulator. Recent studies have shown that HER, as a new AMPK activator, could induce autophagy death in drug-resistant cancers [17]. Furthermore, HER could effectively inhibit the transport function of ABCB1 relative to MDR-linked ABC drug transporters ABCC1 and ABCG2 to enhance the drug-induced apoptosis of tumor cells [18].

In summary, HER has a selective inhibitory effect on the MDR of DOX, which could make DOX better in the treatment of cancer [17,18]. Therefore, it is of great significance to study the pharmacokinetic characteristics of the two drugs in combination. The purpose of this study was to investigate the interaction between HER and DOX on the pharmacokinetics: whether HER could improve the absorption of DOX or the effect in turn, and whether HER could reduce the accumulation of DOX in myocardial tissue.

At present, there are several analytical methods for determining DOX [19,20], but only one article [21] for HER for the pharmacokinetics of rats by LC-MS. According to the previous investigations, the sample preparation methods adopted are mainly focused on a single protein precipitation step. However, there have been no reports of the simultaneous determination of DOX and HER in rat plasma. Therefore, we developed and verified a simple, specific, and sensitive LC-MS/MS method for the simultaneous determination of DOX and HER in rat plasma, and applied it to the pharmacokinetic study of rats to evaluate the effect of HER and DOX interaction on pharmacokinetics.

2. Results and Discussion

2.1. Method Development

To optimize the MS conditions for detecting DOX, HER, and tetrandrine (IS), all the operational parameters were carefully optimized. The analysis showed that positive ion detection has a stronger response than negative ion detection. The MS/MS ion transition was monitored in the MRM mode to improve the specificity and sensitivity of the detection method considering the complexity of biological samples. DOX has the strongest peak at *m/z* 544.2→379.1, HER has the strongest peak at *m/z* 653.4→411.2, and IS (tetrandrine) has the strongest peak at *m/z* 623.3→381.3. The structures of the proposed daughter ions were given by referring to the related articles [19,20,22]. The ion spectra and chemical structures of DOX, HER, and IS are shown in Figure 1.

The chromatographic conditions were optimized, and a good separation effect was obtained with a sharp peak shape, high response, and short run time. The stationary phase and the composition of the mobile phase was studied. An ACQUITY UPLC BEH C_{18} Column (100 mm × 2.1 mm, 1.7 μm) was chosen in this study with good peak symmetry. Different mobile phases (acetonitrile–water and methanol–water or with different concentrations of formic acid or ammonium acetate) were investigated. The results showed that the peak symmetry and response of the acetonitrile–water system were better than those of the methanol–water system. Meanwhile, both gradient elution and isocratic elution were tested, and the result showed that isocratic elution way was more simple, fast, and did not sacrifice any sensitivity and specificity. The retention times of DOX, HER, and IS were 1.46 min, 4.37 min, and 3.65 min, respectively (Figure 2), and the total chromatographic run time was 5.0 min.

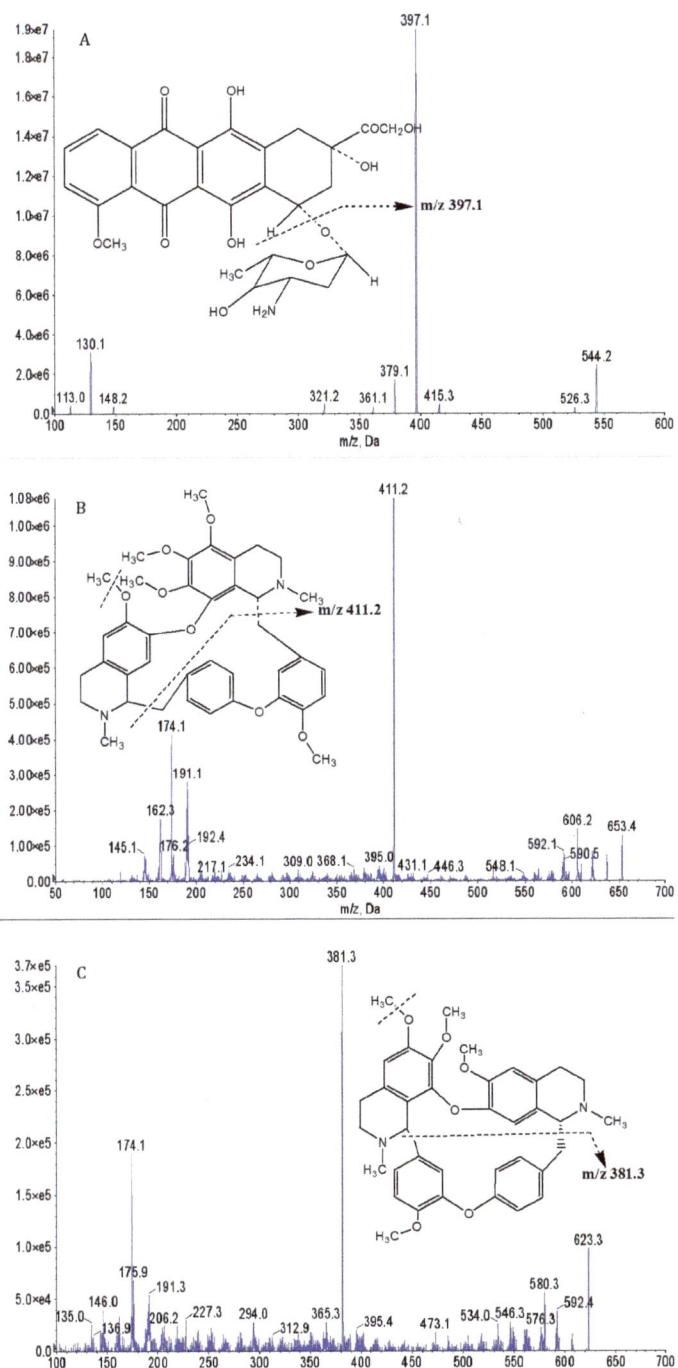

Figure 1. Representative MS of (**A**) doxorubicin (DOX), (**B**) hernandezine (HER), and (**C**) tetrandrine (IS).

ammonium acetate) were investigated. The results showed that the peak symmetry and response of the acetonitrile–water system were better than those of the methanol–water system. Meanwhile, both gradient elution and isocratic elution were tested, and the result showed that isocratic elution way was more simple, fast, and did not sacrifice any sensitivity and specificity. The retention times of DOX, HER, and IS were 1.46 min, 4.37 min, and 3.65 min, respectively (Figure 2), and the total chromatographic run time was 5.0 min.

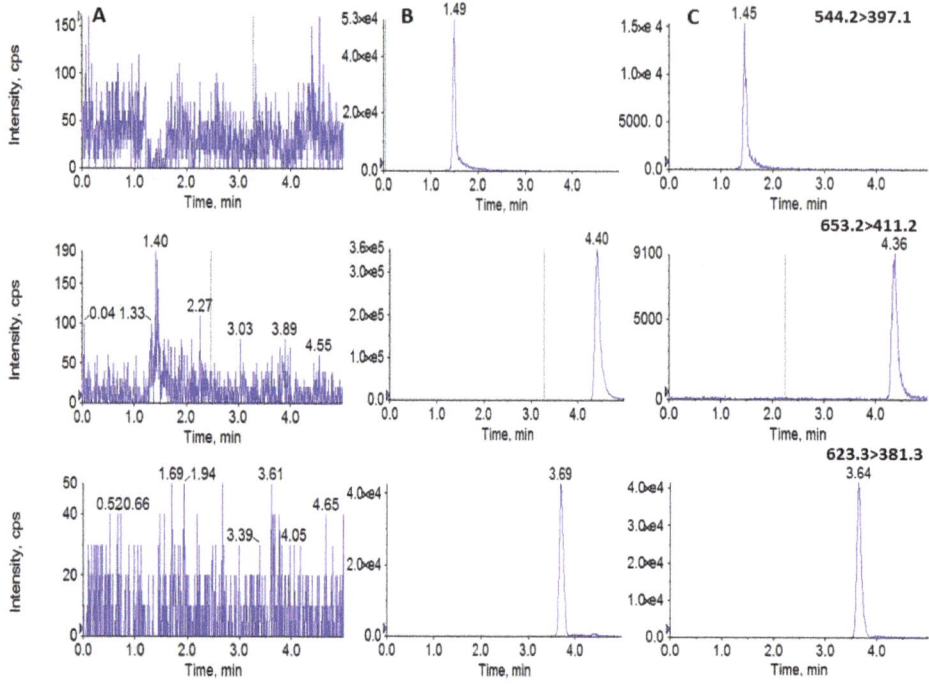

Figure 2. Representative EIC of (**A**) blank plasma; (**B**) blank plasma spiked with doxorubicin, hernandezine at limit of quantification (LOQ) and IS; (**C**) plasma sample after combination administration of the doxorubicin (5 mg/kg) and hernandezine (5 mg/kg) for 2 h.

In this study, a protein precipitation method was firstly considered to prepare samples, which was simple, accurate, and efficient. The extraction recovery and matrix effect were tested. Different precipitation reagents, such as acetonitrile, methanol, and acetonitrile with 0.1% formic acid, were investigated. The results showed that acetonitrile was the best choice, with a higher extraction rate and lower background interference.

2.2. Method Validation

The method validation was conducted in strict accordance with US Food and Drug Administration (FDA) guidelines [23], the content of which consists of selectivity and specificity, linearity, limit of quantification (LOQ), limit of detection (LOD), accuracy, precision, recovery, matrix effect, and stability.

In the process of method development, a selectivity and specificity test was used to verify that the measured substance is the intended analyte to minimize or avoid interference. The selectivity and specificity test of this experiment was demonstrated by the analysis of blank plasma from six individual rats, which was examined by comparing the retention times of DOX, HER, and IS in blank plasma, the addition of DOX and HER at LOQ and IS to blank plasma, and a plasma sample 2 h after an intravenous administration of the mixture of DOX and HER. The blank plasma should be free of interference at the retention times of the analytes and the IS, and which in spiked samples and actual samples should be consistent, respectively. A typical MRM chromatogram of mixed blank plasma in rats, spiked plasma samples with DOX and HER at LOQ and the IS, and plasma samples of rats after an intravenous injection of a mixture of DOX (5 mg/kg) and HER (5 mg/kg) for 2 h is shown in Figure 2. The results showed that there was no significant endogenous interference in the retention time of the analyte under the established chromatographic conditions.

Calibration curves were established by plotting the peak-area ratio (y) between analytes (DOX or HER) and IS against the nominal concentrations. Linearity was evaluated by weighted ($1/x^2$) least squares linear regression analysis. The correlation coefficient (r) should be greater than 0.99, indicating a good linearity. The limit of detection (LOD) is defined as the lowest detectable concentration, judged by the signal-to-noise ratio (SNR) >10. The limit of quantification (LOQ) was defined as the lowest concentration on the calibration curve, which represents the sensitivity of the method and should be lower than the minimum concentration in all the samples. The linear calibration curve was obtained by plotting the peak area ratio (analytes/IS) versus DOX and HER concentration. A weighted ($1/x$) quadratic least-square regression analysis gave typical regression curves. The calibration curves, correlation coefficients, detection ranges, and LOQ of DOX and HER in plasma and myocardial tissues are shown in Table 1. The calibration curves had good linearity with the corresponding range of DOX and HER ($r > 0.99$). Under the optimized conditions, DOX LOQ was <4.0 ng/mL, and HER LOQ was <2.0 ng/mL in rat plasma, judging from the signal-to-noise ratios (SNR) of >10.

Table 1. Calibration curves of doxorubicin and hernandezine in plasma and myocardial tissue homogenate of rats.

Analytes	Samples	Calibration Curves	Correlation Coefficients (r)	Linear Ranges (ng/mL)	LOQs (ng/mL)
DOX	Plasma	Y = 0.0047 + 0.00015x	0.994	32–8000	32
	Heart	Y = −0.0028 + 0.00019x	0.992	32–8000	32
HER	Plasma	Y = −0.0038 + 0.00298x	0.998	20–4000	20
	Heart	Y = −0.0197 + 0.00373x	0.994	20–4000	20

Accuracy and precision tests are critical in determining whether the method is ready for validation, and involve analyzing replicate quality controls (QCs) at different concentrations throughout the assay range. Specifically, the intraday and interday precisions and accuracies were obtained by analyzing five replicates of QC samples at three levels for three consecutive days. Precision, defined as the relative standard deviation (RSD), should be within 15% at each QC level. Accuracy expressed as relative error (RE) must be within ± 15%. Except for the LOQ level, the RSD value of precision should be within 20%, and the RE value of accuracy should be within ± 20%. The intraday and interday precision of the QC samples of DOX and HER were lower than 9.3% and 5.6%, respectively. The accuracy of DOX was −14.0% to 5.5%, and the accuracy of HER was −9.0% to −0.8% (see Table 2). All the assay values were within the range of acceptable variables, indicating that the established method was precise and accurate.

Table 2. Precision and accuracy of doxorubicin and hernandezine in plasma of rats (n = 5). RSD: relative standard deviation.

Analytes	QC Conc. (ng/mL)	Intraday		Interday	
		Precision (RSD, %)	Accurary (mean %)	Precision (RSD, %)	Accurary (mean %)
DOX	80	1.6	−5.7	4.3	−6.0
	800	6.4	5.5	6.8	5.3
	4000	4.0	−14.0	9.3	−3.6
HER	40	5.6	−5.2	4.6	−3.3
	400	5.6	−0.8	3.1	−1.4
	3200	1.4	−7.8	1.9	−9.0

Recovery of the analytes should be optimized to ensure that the extraction is efficient and reproducible. Recoveries of the analytes at three QC levels ($n = 5$) were determined by comparing the peak area ratios of the analytes to IS from QC samples with those of analyte solutions spiked with post-extracted matrix at equivalent concentrations. The matrix effect was examined to assess the possibility of ion suppression or enhancement. The matrix effect was measured by comparing the peak area ratios of the analytes to IS in solutions spiked with the blank processed matrix with the solutions at three QC levels. In common, it was considered that the matrix effect was obvious if the ratio was less than 85% or more than 115%. The recovery and matrix effect data of DOX and HER in rat plasma were shown in Table 3. The matrix effect range of all analytes was 92.9 ± 4.3% to 112.8 ± 1.8%, and the RSD value was lower than 11.6%. The average recovery of DOX and HER at three QC levels was 88.7 ± 6.2% to 108.4 ± 4.9%, and the RSD value was lower than 7.0%. The results showed that this method had no matrix effect, and could be used for biological analysis.

Table 3. Matrix effect and recovery of doxorubicin and hernandezine in plasma of rats (n = 5).

Analytes	QC Conc. (ng/mL)	Matrix Effect		Recovery	
		Mean ± SD (%)	RSD (%)	Mean ± SD (%)	RSD (%)
DOX	80	112.8 ± 1.8	1.6	88.7 ± 6.2	7.0
	800	95.3 ± 11.1	11.6	103.2 ± 2.6	2.5
	4000	92.9 ± 4.3	4.6	95.1 ± 2.1	2.2
HER	40	104.0 ± 1.7	1.7	91.7 ± 4.3	4.7
	400	94.2 ± 1.5	1.6	108.4 ± 4.9	4.6
	3200	94.5 ± 1.7	1.8	93.2 ± 0.6	0.6

Stability was conducted by analyzing three replicates of the samples at three QC levels under the following conditions, including bench top stability after 4 h of exposure at room temperature, auto-sampler stability after 24 h of storage in the auto-sampler at 4 °C, freeze/thaw stability evaluated for three freeze–thaw cycles after freezing at −80 °C and thawing at room temperature, and long-term stability storage at −80 °C for 30 days. The samples were considered stable if the average percentage concentration deviation (expressed as RSD) was within 15% of the actual value. The stability results are shown in Table 4. The variation of all the stability studies was less than 15.0%, which met the standard of stability measurement. Therefore, this method could be used for routine analysis.

Table 4. Stability results for doxorubicin and hernandezine in plasma of rats under different storage conditions (n = 3).

Analytes	QC Conc. (ng/mL)	Bench Top Stability (at Room Temperature for 4 h)	Auto-Sampler Stability (at 4 °C for 24 h)	Freeze/Thraw Stability	Long Term Stability (at −80 °C for 30 days)
DOX	80	1.6	4.3	3.4	3.1
	800	6.4	6.8	6.1	9.6
	4000	4.0	0.2	8.2	5.0
HER	40	5.5	4.7	4.1	1.7
	400	5.5	3.1	1.6	0.8
	3200	1.3	1.9	1.9	1.6

2.3. Pharmacokinetics

This validated method has been successfully applied to the determination of plasma concentration of DOX and HER in rats. In this study, we compared the pharmacokinetic parameters of DOX in the combined treatment group with those in the single treatment group. The pharmacokinetic profiles of HER were also compared in the same way. The mean plasma concentration–time profiles of DOX and HER for the three groups were shown in Figure 3. The pharmacokinetic parameters of DOX and HER in rats following the intravenous administration of single DOX (5 mg/kg), single HER (5 mg/kg), and a combination of DOX and HER (5 mg/kg, respectively) were shown in Table 5.

of DOX in the combined treatment group with those in the single treatment group. The pharmacokinetic profiles of HER were also compared in the same way. The mean plasma concentration–time profiles of DOX and HER for the three groups were shown in Figure 3. The pharmacokinetic parameters of DOX and HER in rats following the intravenous administration of single DOX (5 mg/kg), single HER (5 mg/kg), and a combination of DOX and HER (5 mg/kg, respectively) were shown in Table 5.

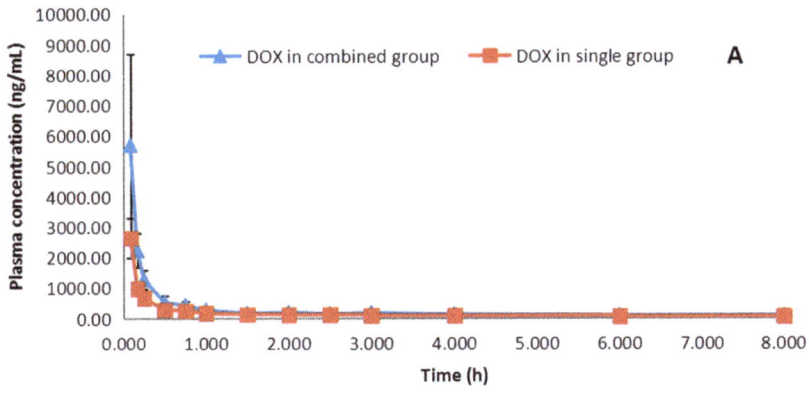

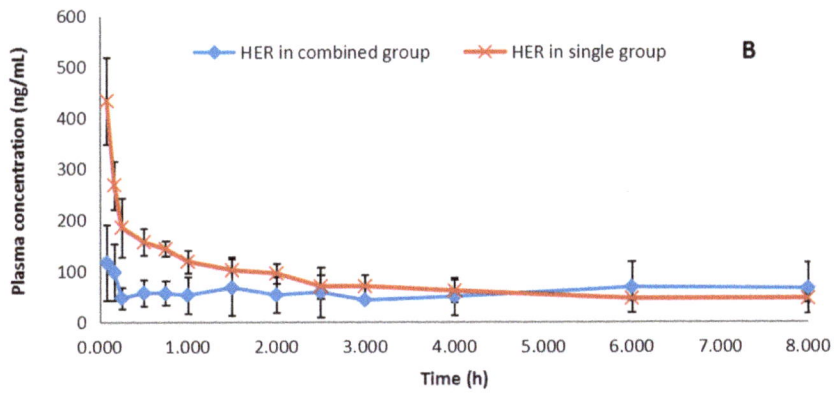

Figure 3. (A) Mean plasma concentration–time curves of doxorubicin in a single doxorubicin group and combination group; (B) Mean plasma concentration–time curves of hernandezine in a single hernandezine group and combination group.

Table 5. Non-compartmental pharmacokinetic parameters of hernandezine and doxorubicin in a single doxorubicin group, single hernandezine group, and combination group (n = 6).

Pharmacokinetic Parameters	Single DOX Group	Single HER Group	Combination Group	
			DOX	HER
C_{max} (ng/mL)	2647 ± 650	433.6 ± 85.2	5703 ± 2980	116.6 ± 74.0
T_{max} (h)	0.083	0.083	0.083	0.083
K_e (1/min)	0.150 ± 0.03	0.090 ± 0.01	0.164 ± 0.02	-
$t_{1/2}$ (h)	4.6 ± 0.8	7.7 ± 1.2	4.2 ± 0.6	-
AUC_{0-t} (ng h/mL)	1109 ± 102	647.2 ± 54.9	1965 ± 142.5	49.9 ± 12.5
$AUC_{0-\infty}$ (ng h/mL)	1412 ± 114	1154 ± 85	2453 ± 218	-
$MRT_{0-\infty}$ (h)	4.9 ± 0.9	10.1 ± 1.4	4.5 ± 0.5	-
CL/F (L/kg/h)	3.5 ± 0.5	4.3 ± 0.5	2.0 ± 0.2	-
Vd/F (L/kg)	23.6 ± 7.1	48.3 ± 5.6	12.4 ± 2.1	81.8 ± 2.3

The C_{max} of DOX in the single group and combined group was 2647 ± 650 ng/mL and 5703 ± 2980 ng/mL, respectively. Meanwhile, the $AUC_{0-\infty}$ was 1412 ± 114 ng/mL and 2453 ± 218 ng/mL, respectively. Meanwhile, the $t_{1/2}$ was 4.6 ± 0.8 h and 4.2 ± 0.6 h, and the $MRT_{0-\infty}$ was 4.9 ± 0.9 h and 4.5 ± 0.5 h, respectively. Significant differences of C_{max} and $AUC_{0-\infty}$ of DOX were observed between the single and combined groups with equivalent doses of DOX administration, which indicated that HER could increase the absorption of DOX. However, there was no significant difference between the $t_{1/2}$ and $MRT_{0-\infty}$ of DOX, which indicated that HER had no effect on DOX's elimination and excretion. In turn, we could see from the plasma concentration–time curves of HER in two treatment groups in Figure 3 that the combination use of DOX made the pharmacokinetic behavior of HER no longer fitted to a non-compartmental model that was used to calculate the pharmacokinetic characteristics in this study. However, we were still able to reach a conclusion from the plasma concentration–time curve and the pharmacokinetic characteristic of HER that the free drug concentration of HER was reduced by the combination use of DOX. The possible reason might be the enhancement of DOX on the drug–protein binding of HER.

The comparison of the accumulated concentrations of DOX in myocardial tissue 8 h after intravenous administration of single DOX and combination of DOX and HER was investigated as shown in Figure 4. A significant difference between the two groups could be observed ($p < 0.05$), indicating that HER was able to reduce the accumulation of DOX in myocardial tissue. Meanwhile, recent studies demonstrated that doxorubicinol (DOX-ol), a secondary alcohol metabolite of DOX [24,25], which may have caused cardiac toxicity by being poorly cleared from the heart and accumulating there to form a long-lived toxicant to heart [26], was to blame. Therefore, the next step is to study whether HER could inhibit the conversion of DOX into DOX-ol, which might be considered as a therapeutic target for DOX-induced cardiac toxicity.

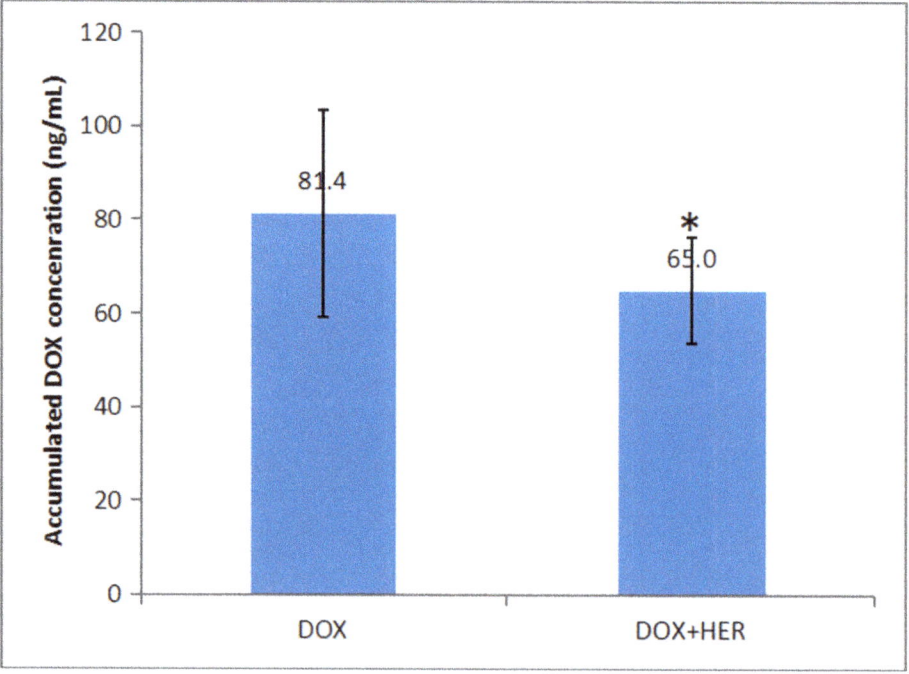

Figure 4. The comparison of the accumulated concentrations of doxorubicin in myocardial tissues 8 h after the intravenous administration of doxorubicin and doxorubicin + hernandezine (mean ± SD, $n = 6$, $p < 0.05$).

3. Experimental

3.1. Chemicals and Reagents

DOX (purity over 99%) was obtained from Dalian Meilun Biotech Co., Ltd. (Dalian, China). HER and tetrandrine (purity over 99%) were purchased from Chengdu Biopurity Phytochemicals Ltd. (Chengdu, China). Ammonium acetate, HPLC-grade, was purchased from Dikma Company (Lake Forest, CA 92630, USA). Acetonitrile and methanol, LC-MS-grade, were purchased from

3. Experimental

3.1. Chemicals and Reagents

DOX (purity over 99%) was obtained from Dalian Meilun Biotech Co., Ltd. (Dalian, China). HER and tetrandrine (purity over 99%) were purchased from Chengdu Biopurity Phytochemicals Ltd. (Chengdu, China). Ammonium acetate, HPLC-grade, was purchased from Dikma Company (Lake Forest, CA 92630, USA). Acetonitrile and methanol, LC-MS-grade, were purchased from Merck KGaA Company (Darmstadt, Germany). Ultra-pure water was provied using a Millipore Milli-Q system (Millipore, Bedford, MA, USA). Other chemical reagents were of analytical grade.

3.2. Animals

Sprague–Dawley rats (male, 250 ± 20 g) were supplied by the Experimental Animal Research Center, China Medical University, China. The rats were raised in a temperature-controlled room at 24 ± 2 °C for free feeding and water intake, and the light/dark cycle was 12 h. The rats were fed for 2 weeks to adapt to the laboratory environment. All the rats fasted for 12 h before the experiment, but with water supplied freely. The protocol for animal care and use in our study (protocol number # CMU2019194) was approved by the Institutional Animal Care and Use Committee at China Medical University.

Eighteen rats were randomly divided into three groups (six rats in each group) and given intravenous treatment with different drugs: group A, DOX (5 mg/kg); group B, HER (5 mg/kg); group C, DOX + HER (5.0 mg/kg, respectively). The injection was prepared in normal saline with 0.5% v/v DMSO.

3.3. Instrumentation and Conditions

The biological samples were analyzed with an Agilent series 1290 UHPLC system (Agilent Technologies, Santa Clara, CA, USA), which was coupled to an AB 3500 triple quadrupole mass spectrometer (AB Sciex, Ontario, ON, Canada) with an electrospray ionization (ESI) source. Data acquisition and instrument control were performed using the 1.6.3 version Analyst software package (AB Sciex, ON, Canada).

The separation process was performed on an ACQUITY UPLC BEH C_{18} Column (100 mm × 2.1 mm, 1.7 μm, Agilent Technologies, Santa Clara, CA, USA). The column temperature was set at 40 °C. The mobile phase was composed of acetonitrile and 10 mM ammonium acetate aqueous solution (70:30, v/v) at the flow rate of 0.3 mL/min in an isocratic elution manner. The injection volume was set at 10 μL.

DOX and HER were quantitatively determined with MRM in the positive ion mode. The MS condition was as follows: the ion spray voltage (IS) was set at 5500 V, the turbo spray temperature (TEM) was set at 500 °C, and the nebulizer gas and heater gas were set at 50 and 50 arbitrary units, respectively. The curtain gas (CUR) was kept at 40 arbitrary units, and the interface heater was on. The collision cell exit potential (CXP) and entrance potential (EP) were set at 7.0 V and 10.0 V, respectively. The declustering potentials (DPs) of DOX, HER, and IS were set at 160 V, 218 V, and 87 V; the collision energies (CEs) were 60 eV, 60 eV, and 27 eV, respectively. Nitrogen was used in all cases. The optimization of the MS transitions for quantification were accomplished as DOX m/z 544.2→379.1, HER m/z 653.4→411.2, and IS (tetrandrine) m/z 623.3→381.3, respectively. Moreover, the qualifier ions for DOX, HER, and IS were set at m/z 321.1, m/z 191.1, and m/z 174.1, respectively.

3.4. Preparation of Stock Solutions, Working Solutions, Calibration Standards, and Quality Control Samples

The standard substances of the analytes were accurately weighed and dissolved in methanol to prepare the DOX and HER stock solution with the concentration of 1.0 mg/mL, respectively. The working solution for preparing calibration standards and QC samples was obtained by diluting the stock solution with acetonitrile–water (50:50, v/v). The IS working solution with a concentration of

200 ng/mL was also prepared. The stock solution and working solution were placed under 4 °C dark condition and brought to room temperature before use.

The calibration standards were prepared by spiking 50 μL of rat blank plasma (or blank myocardial tissue homogenate) with 20 μL of the working solution. The concentration of DOX in rat plasma and myocardial tissue homogenate ranged from 32 to 8000 ng/mL, and HER ranged from 20 to 4000 ng/mL. Low, medium, and high quality control (QC) samples were prepared in the same way as above (40.0, 400, and 3200 ng/mL for DOX; 80.0, 800, and 4000 ng/mL for HER) in both rat plasma and myocardial tissue homogenate. Each concentration needed three replicates.

3.5. Sample Preparation

In this study, DOX, HER, and IS were extracted from the biological matrix (plasma and myocardial tissue homogenate) by routine step protein precipitation. Detailed steps were as follows: take 50 μL of biological matrix, add 20 μL of acetonitrile–water (50:50, *v/v*), and 20 μL of the IS solution, and add 200 μL of precipitation reagent acetonitrile, placed in a 1.5-mL EP tube. Vortex for 1 min, followed by centrifuging at 14,000 rpm for 10 min. Transfer 200 μL of supernatant to another clean 1.5-mL EP tube and centrifuge at 14,000 rpm for another 3 min. Then, an aliquot of 10 μL of the supernatant was injected into the LC-MS system for analysis.

3.6. Pharmacokinetic Study

The method was used to determine the concentration–time profiles of DOX and HER in the plasma of rats after the intravenous administration of DOX (5.0 mg/kg), HER (5.0 mg/kg), and the mixture of DOX and HER (5.0 mg/kg, respectively). Blood samples (250 μL) were taken from the orbital vein at 5 min, 10 min, 15 min, 30 min, 45 min, 1 h, 1.5 h, 2 h, 2.5 h, 3 h, 4 h, 6 h, and 8 h, respectively, and were injected into heparinized 1.5-mL EP tubes. Heparin (2 mg/mL blood volume) was used as an anticoagulant for this study, and blood samples were immediately centrifuged at 14,000 rpm for 10 min at room temperature, followed by a supernatant plasma layer collected and stored at −80 °C for analysis.

After the last blood sample was taken, the rats were sacrificed for cervical dislocation. The heart was removed and rinsed with cold saline to remove the superficial blood. Then, it was blotted dry with filter paper and weighed accurately. After that, the heart was homogenized with normal saline to prepare a homogenate (0.2 g/mL). All samples were stored at −80 °C for analysis.

Plasma concentration–time plots were plotted, and the PK parameters were evaluated by means of non-compartmental pharmacokinetic analysis using DAS 3.2.8 pharmacokinetic program [27]. The PK parameters concerned include half-life ($t_{1/2}$), mean residence time (MRT), area under the plasma concentration–time curve (AUC), clearance (CL), etc. Data was expressed as mean ± SD. The pharmacokinetic parameters were compared using Student's t-test. Differences were considered to be significant at a level of $p < 0.05$.

4. Conclusions

An LC-MS method for the simultaneous determination of DOX and HER in rat plasma was established. The method is sensitive, accurate, easy to follow, and suitable for the pharmacokinetic study. This analytical method has been successfully applied to the pharmacokinetic study of DOX and HER in rats.

The results of this study showed that there were significant differences in the pharmacokinetic parameters of DOX and HER after the intravenous administration of a single dose of DOX, single dose of HER, and a combination of the two. This result might help to explain the influence of DOX and HER interaction on pharmacokinetics and provide a basis for guiding clinical medication.

Author Contributions: The experiments were conceived and designed by X.-S.F. The experiments were equally performed by Y.S. and Y.Z., Chromatographic analysis of plasma and biosamples were performed by W.-P.Z., B.-Z.Z., Analysis of pharmacokinetic behavior of alnustone was finished by K.-F.W. All authors read and approved the final manuscript.

Funding: This research was funded by Key Program of the Natural Science Foundation of Liaoning Province of China (No. 20170541027); Liaoning planning Program of philosophy and social science (No. L17BGL034); and Major subject in education research founded by Chinese Medical Association Medical Education Branch and China Association of Higher Education Medical Education Professional Committee (2018A-N19012).

Acknowledgments: We are very grateful to Q.C. and J.C. for making a major contribution to the revision of the article. The authors are grateful to every person who helped in this work.

Conflicts of Interest: The authors declare no conflicts of interest.

References

1. Minotti, G.; Menna, P.; Salvatorelli, E.; Cairo, G.; Gianni, L. Anthracyclines: Molecular advances and pharmacologic developments in antitumor activity and cardiotoxicity. *Pharmacol. Rev.* **2004**, *56*, 185–229. [CrossRef] [PubMed]
2. Chan, S.; Friedrich, K.; Noel, D.; Pinter, T. Propective randomized trial of docetaxel versus doxorubicin in patients with metastatic breast cancer. *J. Clin. Oncol.* **1999**, *17*, 2341–2354. [CrossRef] [PubMed]
3. Duggan, S.T.; Keating, G.M. Pegylated liposomal doxorubicin. *Drugs* **2011**, *71*, 2531–2558. [CrossRef] [PubMed]
4. Barry, E.; Alvarez, J.A.; Scully, R.E.; Miller, T.L.; Lipshultz, S.E. Anthracycline-induced cardiotoxicity: Course, pathophysiology, prevention and management. *Expert Opin. Pharmacother.* **2007**, *8*, 1039–1058. [CrossRef] [PubMed]
5. Gianni, L.; Herman, E.H.; Lipshultz, S.E.; Minotti, G.; Sarvazyan, N.; Sawyer, D.B. Anthracycline cardiotoxicity: From bench to bedside. *J. Clin. Oncol.* **2008**, *26*, 3777–3784. [CrossRef] [PubMed]
6. Menna, P.; Recalcati, S.; Cairo, G.; Minotti, G. An introduction to the metabolic determinants of anthracycline cardiotoxicity. *Cardiovasc. Toxicol.* **2007**, *7*, 80–85. [CrossRef] [PubMed]
7. Takemura, G.; Fujiwara, H. Doxorubicin-induced cardiomyopathy from the cardiotoxic mechanisms to management. *Prog. Cardiovasc. Dis.* **2007**, *49*, 330–352. [CrossRef]
8. Chatterjee, K.; Zhang, J.; Honbo, N.; Karliner, J.S. Doxorubicin cardiomyopathy. *Cardiology* **2010**, *115*, 155–162. [CrossRef]
9. Szakács, G.; Paterson, J.K.; Ludwig, J.A.; Booth-Genthe, C.; Gottesman, M.M. Targeting multidrug resistance in cancer. *Nat. Rev. Drug Discov.* **2006**, *5*, 219–234. [CrossRef]
10. Zhang, Q.L.; Li, Z.; Cong, H.; Li, J.M.; Sheng, H.M. Comparison of effect of hernandezine on the delayed rectifier potassium current of pulmonary artery smooth muscle cell of normal and pulmonary hypertensive rats. *Chin. J. Pharmacol. Toxicol.* **1998**, *12*, 235–236.
11. Xu, P.X.; Sun, S.S. Effect of the Hernandezine on contraction of rat aortic strips. *Pharmacol. Clin. Chin. Mater. Med.* **1998**, *5*, 15–18.
12. Kruger, M.; Boney, R.; Ordoobadi, A.J.; Sommers, T.F.; Trapani, J.G.; Coffin, A.B. Natural Bizbenzoquinoline Derivatives Protect Zebrafish Lateral Line Sensory Hair Cells from Aminoglycoside Toxicity. *Front. Cell Neurosci.* **2016**, *10*, 83. [CrossRef] [PubMed]
13. Ho, L.J.; Chang, D.M.; Lee, T.C.; Chang, M.L.; Lai, J.H. Plant alkaloid tetrandrine downregulates protein kinase C-dependent signaling pathway in T cells. *Eur. J. Pharmacol.* **1999**, *367*, 389–398. [CrossRef]
14. Virginio, C.; Graziani, F.; Terstappen, G.C. Differential inhibition of rat alpha3* and alpha7 nicotinic acetylcholine receptors by tetrandrine and closely related bis-benzylisoquinoline derivatives. *Neurosci. Lett.* **2005**, *381*, 299–304. [CrossRef] [PubMed]
15. Imoto, K.; Takemura, H.; Kwan, C.Y.; Sakano, S.; Kaneko, M.; Ohshika, H. Inhibitory effects of tetrandrine and hernandezine on Ca^{2+} mobilization in rat glioma C6 cells. *Res. Commun. Mol. Pathol. Pharmacol.* **1997**, *95*, 129–146. [PubMed]
16. Leung, Y.M.; Berdik, M.; Kwan, C.Y.; Loh, T.T. Effects of tetrandrine and closely related bis-benzylisoquinoline derivatives on cytosolic Ca^{2+} in human leukaemic HL-60 cells: A structure-activity relationship study. *Clin. Exp. Pharmacol. Physiol.* **1996**, *23*, 653–659. [CrossRef]

17. Law, B.Y.; Mok, S.W.; Chan, W.K.; Xu, S.W.; Wu, A.G.; Yao, X.J.; Wang, J.R.; Liu, L.; Wong, V.K. Hernandezine, a novel AMPK activator induces autophagic cell death in drug-resistant cancers. *Oncotarget* **2016**, *7*, 8090–8104. [CrossRef]
18. HsiaoS, H.; Lu, Y.J.; Yang, C.C.; Tuo, W.C.; Li, Y.Q.; Huang, Y.H.; Hsieh, C.H.; Hung, T.H.; Wu, C.P. Hernandezine, a Bisbenzylisoquinoline Alkaloid with Selective Inhibitory Activity against Multidrug-Resistance-Linked ATP-Binding Cassette Drug Transporter ABCB1. *J. Nat. Prod.* **2016**, *79*, 2135–2142. [CrossRef]
19. Liu, Y.; Yang, Y.; Liu, X.; Jiang, T. Quantification of pegylated liposomal doxorubicin and doxorubicinol in rat plasma by liquid chromatography/electrospray tandem mass spectroscopy: Application to preclinical pharmacokinetic studies. *Talanta* **2008**, *74*, 887–895. [CrossRef]
20. Chin, D.L.; Lum, B.L.; Sikic, B.I.J. Rapid determination of PEGylated liposomal doxorubicin and its major metabolite in human plasma by ultraviolet-visible high-performance liquid chromatography. *Chromatogr. B Analyt.Technol. Biomed. Life Sci.* **2002**, *779*, 259–269. [CrossRef]
21. Song, Y.; Wang, Z.B.; Zhang, B.Z.; Zhang, W.P.; Zhang, Y.J.; Yang, C.J.; Meng, F.H.; Feng, X.S. Determination of a novel anticancer AMPK activator Hernandezine in rat plasma and tissues with a validated UHPLC-MS/MS method: Application to pharmacokinetics and tissue distribution study. *J. Pharm. Biomed. Anal.* **2017**, *141*, 132–139. [CrossRef] [PubMed]
22. Wu, W.N.; Moyer, M.D. API-ionspray MS and MS/MS study on the structural characterization of bisbenzylisoquinoline alkaloids. *J. Pharm. Biomed. Anal.* **2004**, *34*, 53–56. [CrossRef] [PubMed]
23. U.S. Department of Health and Human Services; Food and Drug Administration; Center for Drug Evaluation and Research (CDER); Center for Veterinary Medicine (CVM). *Bioanalytical Method Validation Guidance for Industry*; Food and Drug Administration: Rockville, MD, USA, 2018.
24. Bains, O.S.; Grigliatti, T.A.; Reid, R.E.; Riggs, K.W. Naturally occurring variants of human aldo-keto reductases with reduced in vitro metabolism of daunorubicin and doxorubicin. *J. Pharmacol. Exp. Ther.* **2010**, *335*, 533–545. [CrossRef]
25. Bains, O.S.; Karkling, M.J.; Grigliatti, T.A.; Reid, R.E.; Riggs, K.W. Two nonsynonymous single nucleotide polymorphisms of human carbonyl reductase 1 demonstrate reduced in vitro metabolism of daunorubicin and doxorubicin. *Drug Metab. Dispos.* **2009**, *37*, 1107–1114. [CrossRef] [PubMed]
26. Salvatorelli, E.; Menna, P.; Gonzalez Paz, O.; Surapaneni, S.; Aukerman, S.L.; Chello, M.; Covino, E.; Sung, V.; Minotti, G.J. Pharmacokinetic characterization of amrubicin cardiac safety in an ex vivo human myocardial strip model. II. Amrubicin shows metabolic advantages over doxorubicin and epirubicin. *Pharmacol. Exp. Ther.* **2012**, *341*, 474–483. [CrossRef]
27. Center of Clinical Drug Research, Shanghai University of Traditional Chinese Medicine, Shanghai, People's Republic of China. Drug Analysis System [CP/DK]. Available online: http://www.drugchina.net/# (accessed on 5 September 2019).

Sample Availability: Samples of the compounds doxorubicin, hernandezine and tetrandrine are available from the authors.

© 2019 by the authors. Licensee MDPI, Basel, Switzerland. This article is an open access article distributed under the terms and conditions of the Creative Commons Attribution (CC BY) license (http://creativecommons.org/licenses/by/4.0/).

Article

A Hydrophilic Interaction Liquid Chromatography–Tandem Mass Spectrometry Quantitative Method for Determination of Baricitinib in Plasma, and Its Application in a Pharmacokinetic Study in Rats

Essam Ezzeldin [1,2], Muzaffar Iqbal [1], Yousif A. Asiri [3], Azza A Ali [4], Prawez Alam [5] and Toqa El-Nahhas [4,*]

[1] Department of Pharmaceutical Chemistry and Drug Bioavailability Unit, Central Laboratory, College of Pharmacy, King Saud University, Riyadh 11451, Saudi Arabia; esali@ksu.edu.sa or ezzeldin24@hotmail.com (E.E.); muziqbal@ksu.edu.sa (M.I.)
[2] National Organization for Drug Control and Research, Cairo 12611, Egypt
[3] Clinical Pharmacy Department, College of Pharmacy, King Saud University, Riyadh 11451, Saudi Arabia; yasiri@ksu.edu.sa
[4] Pharmacology and Toxicology Department, Faculty of Pharmacy (Girls) Al-Azhar University, Cairo 11754, Egypt; azzamro@gmail.com
[5] Pharmacognosy Department, College of Pharmacy, Prince Sattam Bin Abdulaziz University, Al-Kharj 11942, Saudi Arabia; p.alam@psau.edu.sa
* Correspondence: toqa_elnahhas1@yahoo.com; Tel.: +201012230599

Academic Editors: Constantinos K. Zacharis and Aikaterini Markopoulou
Received: 29 February 2020; Accepted: 28 March 2020; Published: 31 March 2020

Abstract: Baricitinib, is a selective and reversible Janus kinase inhibitor, is commonly used to treat adult patients with moderately to severely active rheumatoid arthritis (RA). A fast, reproducible and sensitive method of liquid chromatography-tandem mass spectrometry (LC-MS/MS) for the quantification of baricitinib in rat plasma has been developed. Irbersartan was used as the internal standard (IS). Baracitinib and IS were extracted from plasma by liquid–liquid extraction using a mixture of n-hexane and dichloromethane (1:1) as extracting agent. Chromatographic separation was performed using Acquity UPLC HILIC BEH 1.7 µm 2.1 × 50 mm column with the mobile phase consisting of 0.1% formic acid in acetonitrile and 20 mM ammonium acetate (pH 3) (97:3). The electrospray ionization in the positive-mode was used for sample ionization in the multiple reaction monitoring mode. Baricitinib and the IS were quantified using precursor-to-production transitions of m/z 372.15 > 251.24 and 429.69 > 207.35 for baricitinib and IS, respectively. The method was validated according to the recent FDA and EMA guidelines for bioanalytical method validation. The lower limit of quantification was 0.2 ng/mL, whereas the intra-day and inter-day accuracies of quality control (QCs) samples were ranged between 85.31% to 89.97% and 87.50% to 88.33%, respectively. Linearity, recovery, precision, and stability parameters were found to be within the acceptable range. The method was applied successfully applied in pilot pharmacokinetic studies.

Keywords: baricitinib; UPLC-MS/MS; pharmacokinetic study; irbersartan

1. Introduction

Rheumatoid arthritis (RA) is a progressive and long-lasting autoimmune disorder that most commonly affects the joints. RA is a chronic systemic inflammatory disorder which affects 0.5–1% of the adult population. It is characterized by persistent inflammation of synovial joints resulting symmetrical inflammatory polyarthritis with progressive erosive destruction of joints including cartilage and bone deformity and damage [?]. The main symptoms of RA are joint pain, inflammation, swelling, stiffness and generalized fatigue with episodes of RA flares and remissions. RA is a systemic disease that can have a significant effect on other organs, including eyes [?], heart [??], and lung [?].

Janus kinase inhibitors are anti-rheumatic drugs that target the intracellular kinase (JAK). JAKs play an important role in the signaling of many cytokines and hematopoietic growth and development [?]. Baricitinib (Figure 1A), a selective and reversible JAK1 and JAK2 inhibitor, is commonly used to treat moderate to severe active RA in adult patients with poor response to one or more tumor necrosis factor inhibitors (TNFis) [?]. Baricitinib has showed dose-dependent efficacy for up to 24 weeks with significant improvements in the signs and symptoms of RA in patients [?]. It has been reported that patients treated with baricitinib demonstrated consistent pain relief irrespective of the degree of inflammation control. Baricitinib treatment is associated with improvements in RA disease control, provides rapid pain relief, improves work productivity and is cost-effective in comparison to other RA drugs [?]. In addition, patients with hereditary immunodeficiency syndrome showed a complete disappearance of rheumatoid nodules following 12 months of baricitinib treatment. Moreover, baricitinib had a significant improvement in the signs and symptoms of systemic lupus erythematosus [?].

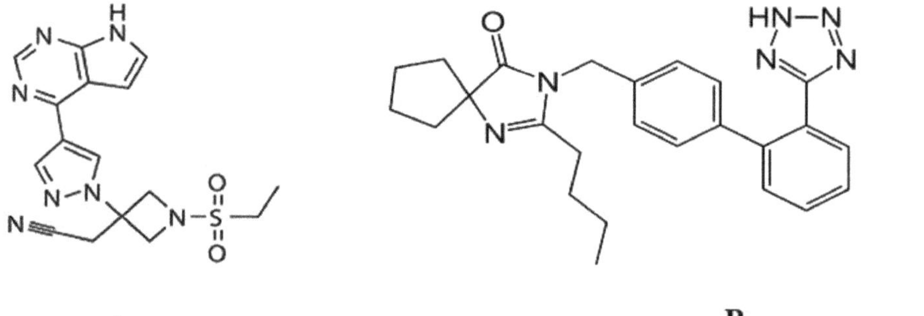

Figure 1. Chemical structures of baricitinib (**A**) and irbesartan (**B**).

Baricitinib is rapidly absorbed after oral administration and reaches peak plasma concentrations at approximately 1 h. Baricitinib oral bioavailability is 79% and it is moderately bound (about 50%) to plasma proteins mainly albumin [?]. Baricitinib is primarily metabolized by CYP3A4 enzyme and it is a substrate of P-glycoprotein (Pgp), organic anionic transporter (OAT) 3, multidrug and toxic extrusion protein (MATE) 2-K and breast cancer resistance protein (BCRP). Baricitinib pharmacokinetics was found to be affected by food and concurrent administration of other drugs. The therapeutic effects and safety of baricitinib are related to the rate and extent of absorption [?]. It was also found that probenecid elevates the extent of baricitinib absorption, which represent in the potentiation of AUC also by twofold and decreased the renal clearance to 69% while the rate of absorption (C_{max}) was not significantly increased [?]. Furthermore, rifampicin, a potent CYP3A inducer, decreased baricitinib AUC without affecting C_{max} [?]. No clinically significant drug interaction has been reported between CYP substrates, inhibitors and inducers with baricitinib. Ketoconazole (CYP3A inhibitor) and fluconazole (CYP2C19/CYP2C9/CYP3A inhibitor) did not significantly affect baricitinib pharmacokinetics, however baricitinib clearance was reduced by cyclosporin A, an inhibitor of the transporter OAT3. Although co-administration of high-fat meal had no effect on baricitinib AUC, it reduced C_{max} by about 29% with 3 h delayed t_{max} and had no clinical influence [?].

Baricitinib safety is a controversial area. Some studies showed that baricitinib was safe and well tolerated following treatment of patients with moderate to severe RA for 24 weeks [???]. However, other studies demonstrated that baricitinib induces a stable dose-response increase in LDL-C and HDL-C levels with developing the risk of thrombosis [?], elevates infection risk, particularly for herpes zoster [?] and GI perforations [?]. Substantial data concerning baricitinib embryotoxicity and teratogenicity in humans are lacking. However, pre-clinical animal studies revealed its embryotoxicity, teratogenicity, and adverse effect on bone development in utero at higher dosages [?].

RA therapy includes antirheumatic drugs, JAK inhibitors, biologic (such as tumor necrosis factor-α or and interleukin 6 inhibitors), and nonsteroidal anti-inflammatory drugs mostly in combinations [?]. Therefore, it is important to study baricitinib pharmacokinetics in different therapeutic approaches. Currently, TKIs including baricitinib are mostly used as a fixed dosage therapy with no consideration of inter-individual differences. Although this might be a practical tool, optimal target therapeutic plasma concentration can be achieved based on dosage individualization strategy using therapeutic drug monitoring to obtain rapid and effective clinical response with minimum incidence of adverse effects [?].

There are very limited resources for validated analytical methods of baricitinib in biological fluids. Both available methods measure baricitinib in combination with other drugs using either low purification technique, protein precipitation with low sensitivity and long run time [?] or solid phase extraction which is an expensive technique and required high volume of samples [?]. Hence, development of a fast and sensitive method for analysis of baricitinib in biological fluids is of great importance for better understanding of baricitinib performance in both routine therapeutic drug monitoring and clinical trials. The present study aimed to develop a new, rapid, sensitive, reproducible and cost-effective method for determination of baricitinib in plasma and to study its applicability in a pharmacokinetic study in rats.

2. Results and Discussion

2.1. Chromatography and Mass Spectrometry Conditions

Mass spectrometry parameters were optimized to achieve better ionization of baricitinib and IS molecules. MS optimization was obtained by direct injection of baricitinib and IS into the mass spectrometer. The mass spectra revealed that more stable and higher responses were obtained in positive-ion mode. Cone voltage and collision energy parameters were also optimized. The mass spectra revealed peaks for the most abundant protonated molecular ions $[M + H]^+$ for baricitinib and IS were obtained at m/z 372.36 and 429.69, respectively. The major (predominant) fragment ions observed in each product spectrum were at m/z 251.24 and 207.15 for baricitinib and IS, respectively (Figure ??). Irbersartan was used as internal standard as it has some physical properties similar to the anayte such as poor solubility in buffer solutions which is essential for separation on HILIC column and was efficiently extracted with the same solvent. Moreover, it was relatively separated at the same retention time of the analyte which helps to decrease the separation run time.

Different types of column were tested and Acquity UPLC HILIC BEH 1.7 μm 2.1 × 50 mm showed good separation. HILIC chromatographic separation mechanism depends on physicochemical properties of the stationary phase and mobile phase. The mobile phase for HILIC chromatography includes water-miscible polar organic solvents with a small amount of water or buffer [?]. We attempted several organic solvents to use for the mobile phase. Different strengths of acetate and format buffer concentrations, as well as different pH levels were tested in order to achieve the optimal separation of baricitinib and IS. Due to the nature of acetonitrile as a water-miscible organic solvent with intermediate polarity and lacking an acidic proton characteristic that encourages retention of polar analytes and the ability of ammonium acetate to control pH of the mobile phase and ion strength, they were selected for use as the mobile phase. 20 mM ammonium acetate buffer solution at pH 3 was used for the mobile phase because it helped to suppress baricitinib peak tailing and to obtain symmetric and sharp peaks. Hence, the final choice of the mobile phase was 0.1% formic acid in acetonitrile and 20 mM ammonium acetate buffer (97:3) at pH 3. The flow rate was 0.2 mL/min. The low amount of buffer in the mobile phase is in agreement with Buszewski and Noga [?]. LC-MS/MS method described here showed high sensitivity and has a short run time (3.0 min) which is acceptable for routine analyses, which has advantages over the previous method described by Veeraraghavan et al. [?].

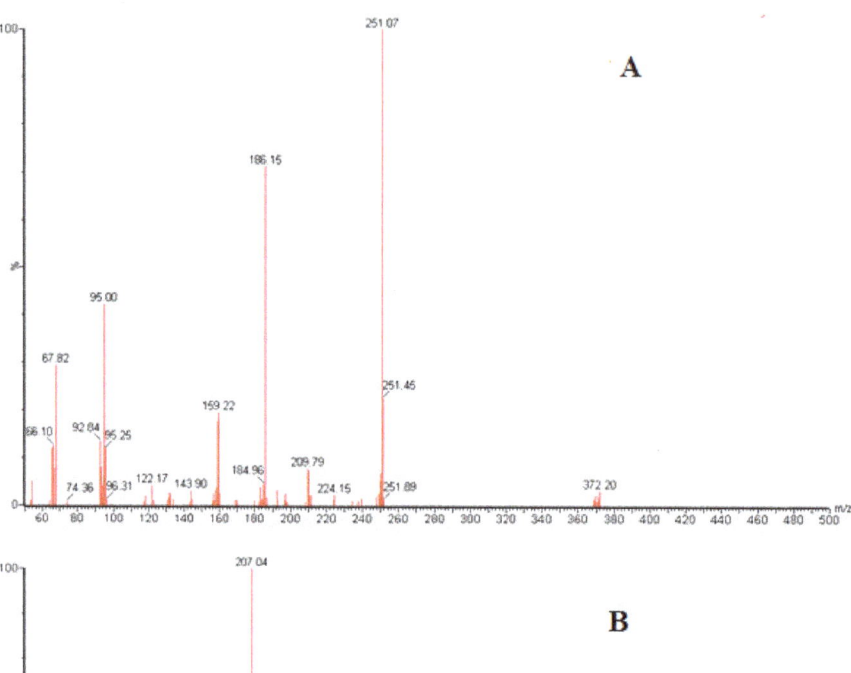

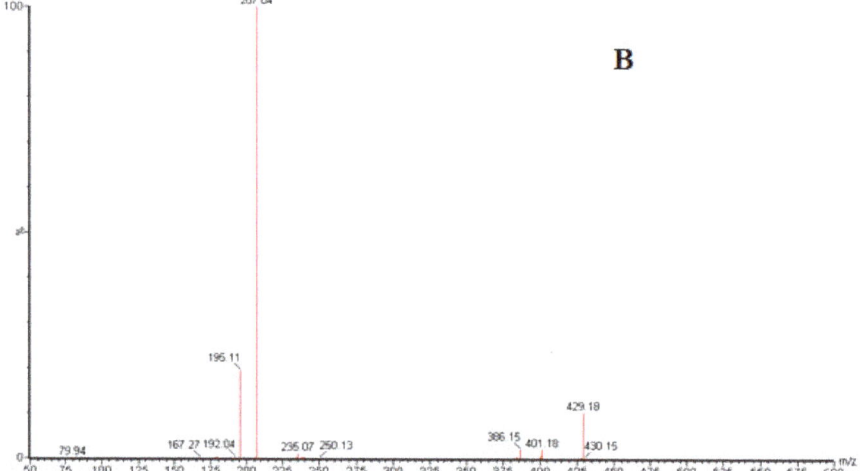

Figure 2. Positive ESI product ion mass spectra of baricitinib (**A**) and irbersartan (**B**) obtained from collisional activated dissociation of the precursor ion m/z 372.20 and 429.18, respectively.

Different types of column were tested and Acquity UPLC HILIC BEH 1.7 μm 2.1 × 50 mm showed good separation. HILIC chromatographic separation mechanism depends on physicochemical properties of the stationary phase and mobile phase. The mobile phase for HILIC chromatography includes water-miscible polar organic solvents with a small amount of water or buffer [14]. The optimal separation of baricitinib and IS. Due to the nature of acetonitrile as a water-miscible organic solvent with intermediate polarity and lacking an acidic proton characteristic that encourages the optimal separation of baricitinib and IS. Due to the nature of acetonitrile as a water-miscible organic solvent with intermediate polarity and lacking an acidic proton characteristic that encourages good retention of polar analytes and the ability of ammonium acetate to control pH of the mobile phase

2.2. Method Validation

2.2.1. Selectivity and Specificity

Under the optimized conditions, there were no significant interfering peaks from endogenous sources that could be observed at the retention times of baricitinib and IS in the blank plasma obtained from six different rats. The retention times of baricitinib and IS were 1.2 ± 0.02 and 1.8 ± 0.03 min, respectively, with a total run time of 3.0 min only. Additionally, peaks were detected with excellent resolution and shapes, which concludes acceptable selectivity of the method for routine quantification of baricitinib in plasma samples. Representative MRM chromatograms of the blank plasma did not show any interfering peaks at the elution times of baricitinib and IS, as presented in Figure 3A.

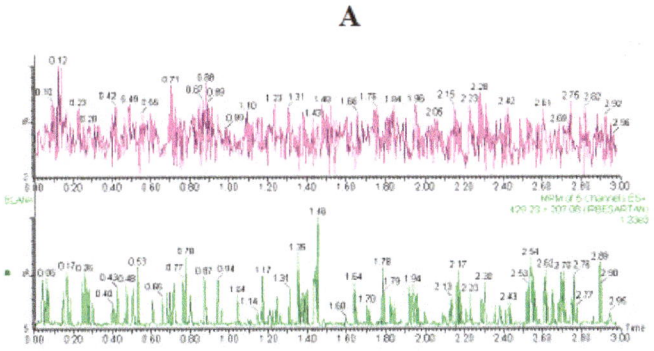

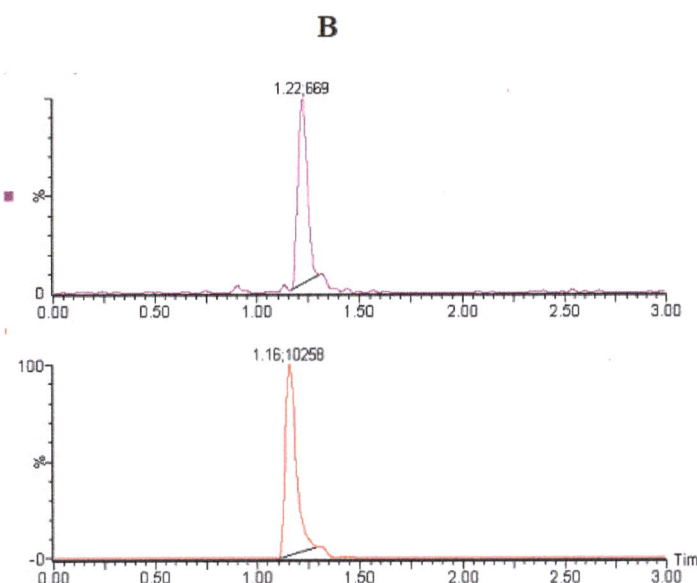

Figure 3. MRM Chromatograms of baricitinib and internal standard in blank rat plasma (**A**), and plasma spiked at LLOQ level (**B**).

2.2.2. Linearity and Lower Limit of Quantification

The chromatogram of baricitinib and IS in blank plasma and LLOQ are presented in Figure ??B. At the LLOQ, the signal to noise ratio was greater than 5-fold of the response of the blank sample. The calibration curves were constructed by plotting peak area ratios (baricitinib/IS) versus concentrations of baricitinib ranging from 0.2 to 500 ng/mL. The weighing factor of 1/X2 was used for the linear fitting and least square residual for the calibration curves. The correlation coefficient was found to be ≥ 0.997, LLOQ was quantified with acceptable accuracy and precision (≤20%) (Table ??).

Table 1. Intra-day and inter-day precision and accuracy values of baracitinib in rat plasma.

Nominal Conc. (ng/mL)	Intra-Day			Inter-Day		
	Measured Conc. (ng/mL)	CV (%)	Accuracy (%)	Measured Conc. (ng/mL)	CV (%)	Accuracy (%)
0.2	0.17 ± 0.02	13.2	85.3	0.175 ± 0.02	11.4	87.5
0.6	0.54 ± 0.06	11.8	89.4	0.53 ± 0.06	11.3	88.3
40.0	35.99 ± 4.00	11.1	90.0	35.24 ± 3.30	9.4	88.1
400.0	344.36 ± 2.49	2.8	86.9	353.16 ± 6.61	1.9	88.3

2.2.3. Precision and Accuracy

The intra- and inter-day precision and accuracy of the method used are listed in Table ??. Notably, the intra-day and inter-day precision values for the QC concentrations were ≤13.2% and ≤11.4 (expressed as CV %), respectively. Similarly, intra-day and inter-day accuracy ranged from 85.3% to 90.0% and 87.5% to 88.3%, respectively. The results showed that the assay met the desired acceptable precision and accuracy criteria set by regulatory guidelines.

2.2.4. Recovery and Matrix Effects

The extraction efficiency (recovery) and matrix effect of baricitinib using QC samples at three different concentrations (0.6, 40.0, 400.0 ng/mL) and IS (100 ng/mL) are presented in Table ??. The average extraction efficiency of baricitinib was 87.9%. This was consistent and concentration-independent of the CV% value was ≤8.2%. The average matrix effect value of baricitinib between three QC concentration levels was 88.8%. The matrix effect was considered negligible as CV% was ≤10.3% for all QC concentration levels.

Table 2. Recovery and matrix effect of baracitinib and IS in rat plasma ($n = 6$).

Drug Name	Nominal Conc. (ng/mL)	Extraction Recovery			Matrix Effects		
		Mean ± SD	Accuracy (%)	CV (%)	Mean ± SD	Accuracy (%)	CV (%)
Baracitinib	0.6	0.58 ± 0.04	95.9	5.4	0.63 ± 0.04	89.3	6.8
	40.0	34.37 ± 1.85	85.9	8.2	0.53 ± 0.04	89.5	6.8
	400.0	351.67 ± 31.32	81.7	0.7	349.54 ± 35.94	87.7	10.3
Irbersartan	100.0	74.33 ± 3.21	74.3	4.3	87.40 ± 2.95	87.4	3.5

2.2.5. Stability

Baricitinib and IS were stable under analysis process and storage conditions. The QC samples analysis showed no significant changes in comparison to nominal concentrations. In all cases, accuracy and precision of QC samples were found to be within the acceptable limits of ±15% against freshly prepared calibration curves as shown in Table ??. Moreover, baricitinib and IS working solutions were stable (with no significant changes) at room temperature (23–25 °C) for 6 h and at refrigerated temperature (2–8 °C) for 10 days.

Table 3. Stability quality control sample of baracitinib in rat plasma ($n = 6$).

Stability	Conc. (ng/mL)					
	40.0 ng/mL			400.0 ng/mL		
Parameters	Mean ± SD	Accuracy (%)	Precision (%CV)	Mean ± SD	Accuracy (%)	Precision (%CV)
Bench top (6 h)	34.61 ± 3.03	8.7	86.5	343.27 ± 44.76	85.8	13.0
Thaw/freeze (3 cycles)	34.09 ± 3.99	85.1	8.8	345.73 ± 39.84	86.4	11.5
Auto-sampler	35.6 ± 2.69	89.0	7.6	360.36 ± 11.52	90.1	11.5
Long term (at −80 °C for 8 weeks)	34.63 ± 4.63	86.6	13.4	350.56 ± 28.53	87.64	8.1

2.3. Application to a Pharmacokinetic Study

To verify the sensitivity, selectivity, and efficiency of this method, the method was applied in a preliminary pharmacokinetic study of baricitinib in male rats after oral administration of 2 mg/kg suspension of baricitinib. The results of the pharmacokinetic parameters are presented in Table ??. The C_{max} of 129.08 ± 91.4 ng/mL was achieved at 0.5 h after administration of 2 mg/kg of baricitinib suspension. AUC_{0-11} and $AUC_{0-\infty}$ were found to be 205.15 ± 101.4 and 222.53 ± 107.2 ng h/mL, respectively. The values of baricitinib pharmacokinetic parameters in the present work are much lower than the values reported by Veeraraghavan et al. [?]. The difference was 5-fold higher C_{max} and 4.3-fold higher AUC. However, the values obtained in the current study is highly comparable to the study reported by Committee for Medicinal Products for Human Use, EMA [?].

Table 4. Pharmacokinetic Parameters (Mean ± SD) of baricitinib Following Administration of 2 mg/kg to rats.

Parameters	Mean * ± SD
C_{max} (ng/mL)	129.08 ± 91.4
AUC_{0-11} (ng.h/mL)	205.15 ± 101.40
AUC_{0-inf} (ng.h/mL)	222.53 ± 107.20
K_{el} (h)	0.32 ± 0.04
$t_{1/2}$ (h)	2.24 ± 0.43
MRT (h)	3.30 ± 0.78
t_{max} (h)	0.5

*: Median for t_{max}. C_{max}, maximum concentration; T_{max}, time to reach maximum concentration; AUC_{0-11}, area under the concentration time curve from 0 to 11 h; AUC_{0-inf}, AUC from 0 h to infinity; $t_{1/2}$, half-life; MRT, median residence time.

Representative MRM chromatograms of baricitinib and IS at 1.0 h after oral administration of baricitinib are shown in Figure ??. Mean plasma concentration versus time profile of baricitinib in rats is also shown in Figure ??. Notably, the LC-MS/MS method satisfied the requirement of routine analyses as it had a short run time (3 min).

The only published method involved protein precipitation procedure with a gradient reversed-phase LC-MS/MS [?]. Although protein precipitation is widely used as a simple and rapid sample preparation technique for bioanalysis, it has some disadvantages. Sample cleaning procedure is relatively poor. The matrix components of extracted samples are not efficiently removed. Therefore, these components may co-elute with the analytes in the isolated supernatant and interfere with the ionization process either by enhancing or suppressing ion production. Moreover, this approach can arise chromatographic separation problems from column fouling or blockage since the efficiency of precipitation is not optimal [? ?]. The present method involved liquid–liquid extraction technique using single step sample extraction procedure. It offers a rapid sample preparation approach without any tedious or time-consuming steps. This method provides efficient peak resolution in isocratic elution mode. There was no need for gradient elution of the analytes as reported by Veeraraghavan et al. [?]. Moreover, the current method is more sensitive and has a wider calibration range than the previously published method. This allows the application of this method not only in pharmacokinetic studies but also in toxicokinetics studies. Moreover, the flow rate of this method is relatively low (0.25 mL/min) which is four times less compared to that of Veeraraghavan et al. [?] method (1.00 mL/min). Furthermore, the run time is short (3 min) with both the analyte and IS eluted at retention time of 1.22 and 1.16 min, respectively. This is shorter run time in comparison with Veeraraghavan et al. method (7.5 min) in which all the analytes of interest eluted between 3.5–5.3 min [?]. Therefore, our method has the advantage of minimizing analysis time, solvent consumption and consequently the relative cost.

Furthermore, the run time is short (3 min) with both the analyte and IS eluted at retention time of 1.22 and 1.16 min, respectively. This is shorter run time in comparison with Veeraraghavan et al. method (7.5 min) in which all the analytes of interest eluted between 3.55-3.3 min [22]. Therefore, our method has the advantage of minimizing analysis time, solvent consumption and consequently the relative costs.

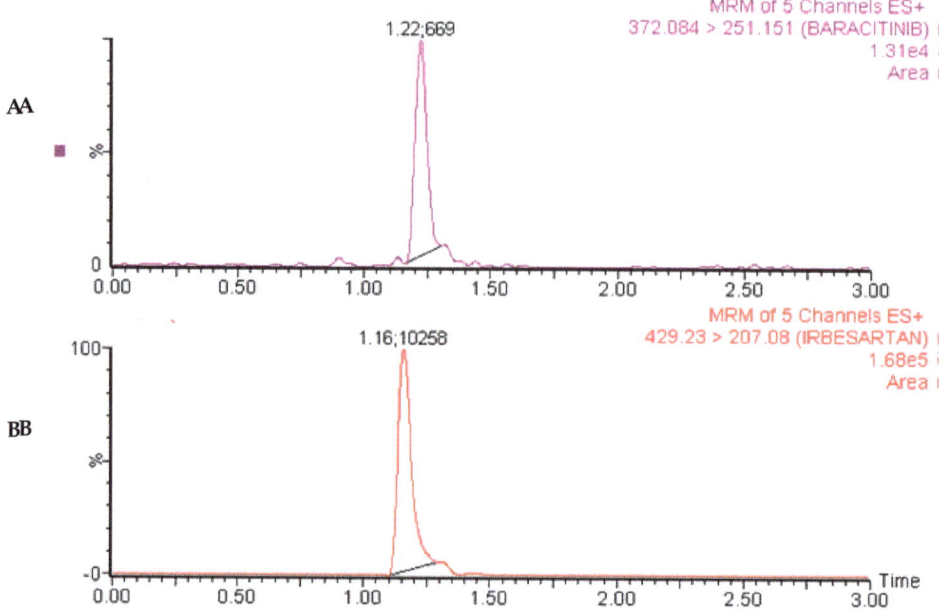

Figure 4. MRM Chromatograms of baricitinib (A) and internal standard (B) in real rat plasma sample obtained at 1.0 h following oral administration of 2 mg/kg baricitinib.

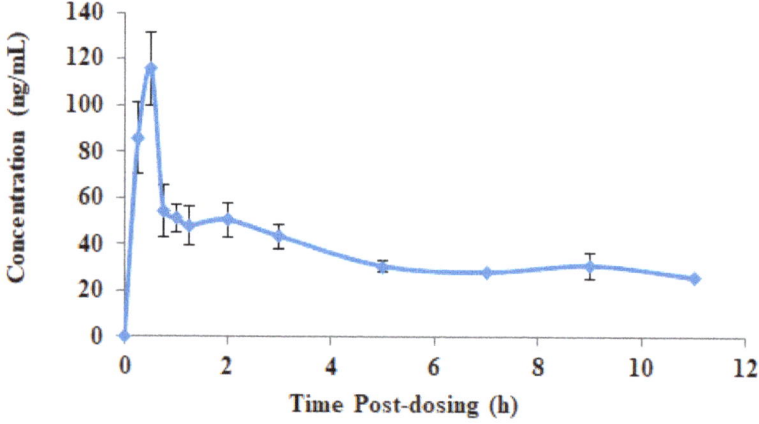

Figure 5. Mean plasma concentration-time profiles of baricitinib.

3. Materials and Methods

3.1. Experimental

Baricitinib (>99% purity) was purchased from Enzo Life Sciences, Inc. (Exeter, UK) and irbersartan (Figure ??B) was purchased from Beijing Mesochem Technology Co., Ltd. (Beijing, China). HPLC-grade methanol and acetonitrile were obtained from Avonchem Ltd., (Macclesfield, UK) and Winlab Pty. Ltd. (Brendale, Australia), respectively. Analytical-grade formic acid, dichloromethane, n-hexane, and ammonium acetate were obtained from the BDH Laboratory (Lutterworth, UK). Blank rat plasma was collected and separated from the blood of healthy rats.

3.2. Equipment

Waters Acquity TQD UPLC/ Mass spectrometer (Waters Co., Milford, MA, USA) was used in the study. Other equipment used included vortex mixer Taboys® model AP 56 (TROEMNER. Hingham, MA, USA), analytical balance Mettler Toledo® model XS 205 (Greifensee, Switzerland), and a sample concentrator Thermo® Savant SC210A speed Vac (Waltham, MA, USA). Deionized water was prepared by Milli-Q reverse osmosis (Millipore®, Bedford, MA, USA).

3.3. Chromatographic Conditions

Sample analysis was performed on an Acquity TQD UPLC-MS/MS system (Waters Co., Milford, MA, USA). The triple quadrupole mass spectrometer was operated in the MRM mode, and detection was carried out using electrospray ionization (ESI) in the positive ion mode. In addition, the mass spectrometry parameters including parent to daughter ion transition, collision energy, cone voltage, and dwell time were optimized to determine the analyte and IS (Table ??). The collision gas (argon) flow rate was kept at 0.1 mL/min and the desolvation gas (nitrogen) flow rate was optimized to 600 L/h. Notably, chromatographic separation was achieved using a Waters Acquity UPLC HILIC BEH 1.7 μm 2.1 × 50 mm column with the mobile phase consisting of 0.1% formic acid in acetonitrile and 20 mM ammonium acetate (pH 3). The flow rate was 0.2 mL/min.

Table 5. Mass optimization parameter for, baracitinib and irbersartan (IS).

Parameters	Baracitinib	Irbersartan
I. Parameters of compound-dependent		
SRM transition (m/z) (Parent)	372.15	429.20
Daughter	251.24	207.35
Collision energy (eV)	52	38
Cone voltages	30	22
II. Parameters of source-dependent		
Collision gas	Argon with a flow rate of 0.1 mL/min	
Desolvating gas	Nitrogen with flow rate of 600 L/h	
Desolvating temperature (°C)	350	
Source temperature was (°C)	150	
The capillary voltage (kV)	4	

3.4. Preparation of Stock and Working Solutions

The standard stock solutions of baricitinib and irbesartan were prepared individually by dissolving 5 mg of each drug in dimethyl sulfoxide (DMSO, Loba Chemi, Mumbai, India). Stock solutions were then diluted with methanol to obtain working solutions of 1000 μg/mL for both baricitinib and IS. Further dilution was carried out in acetonitrile to yield working solutions (20 μg/mL) for both the analyte and IS. The calibration range was 0.2 to 500 ng/mL (0.2, 1.0, 5.0, 1.0, 20.0, 100.0, 200.0, and 500.0 ng/mL). Three quality control samples at three concentration levels 0.6, 40.0, and 400.0 ng/mL and lower limit of quantitation (LLOQ; 0.2 ng/mL) were prepared.

3.5. Sample Preparation

Baricitinib and IS were extracted from rat plasma using liquid–liquid extraction. 20 μL of irbesartan working solution (20 μg/mL) was added to 100 μL plasma samples. After vortex-mixing for 20 s, 1 mL of *n*-hexane:dichlromethane mixture (1:1) was added. Samples were then mixed for 1 min and centrifuged for 5 min at 5600× *g*. The supernatant (0.8 mL) was transferred to a 5-mL tube and evaporated to dryness at 40 °C. Residues were reconstituted in 100 μL of the mobile phase and 5 μL was injected into the LC-MS/MS system.

3.6. Method Validation

The method was validated according to the FDA and EMA guidelines for bioanalytical method validation [? ?].

The selectivity of the proposed method was estimated by analyzing rat blank plasma samples from 6 different sources to detect the interfering peaks at the same retention times of the analyte and IS. The sensitivity of the method is expressed by the LLOQ, which is defined as the lowest concentration of an analyte that can be accurately measured.

The plasma calibration curves were constructed by plotting the peak area ratios (baricitinib/IS) against baricitinib concentrations covering the expected range (0.2–500 ng/mL), including the LLOQ. The correlation coefficient (r^2) of calibration curves should not less than 0.99. The limit of detection (LOD) is defined as the concentration of an analyte yielding a peak with a signal to noise ratio of 3, while this ratio is 5 times for the low limit of quantification (LOQ).

The accuracy and precision of the method were determined by analyzing three different QC concentrations (low, medium and high QC samples) representing the entire range of the calibration curve and LLOQ samples. Inter-day accuracy and precision were measured consecutively for three days, whereas intra-day accuracy and precision were measured in one day. Accuracy of QC samples should be within ±15% of the nominal concentrations of QC samples and 20% for LLOQ. Precision should not exceed 15% CV% for QC samples and 20% for LLOQ.

Stability of the analyte was assessed using five measurements of QC samples at low and high concentrations after exposure to different conditions of storage and processing temperature. Baracitinib stability was evaluated after three freeze-thaw cycles, storage at room temperature (23–25 °C) for 6 h (short term stability), and subsequent storage for eight weeks at −80 °C (long term stability). Moreover, the stability of stock and working solutions of baricitinib and the IS at room temperature for 6 h (23–25 °C) and at refrigerator (2–8 °C) for 12 days was tested.

The absolute recovery of the method was determined by comparing the average of peak area measurements obtained from the blank plasma spiked before extraction to those obtained from the blank plasma spiked after extraction. The effect of the matrix was evaluated by comparing the peak area ratios of post-extracted plasma spiked with baricitinib and IS to peak area ratio of real solutions contain the same concentration.

3.7. Animal

Sprague Dawely rats weighing 200–230 g were obtained from the animal house at the National Organization for Drug Control and Research, Giza, Egypt. Rats were fed standard sufficient food and water and experiments were performed under the Guide for Care and Use of laboratory animals. experimental protocols were approved by the ethics committee of the Faculty of Pharmacy at Al-Azhar University, Cairo, Egypt (no. 206).

3.8. Application to a Pharmacokinetic Study

To demonstrate the utility of the present method, a pharmacokinetic study of baricitinib was performed on six male Sprague Dawely rats (200–230 g). Baricitinib was used to create a suspension using 1% carboxymethyl cellulose. Following overnight fasting, the drug was administered in a dose of 2 mg/10 mL/kg. Blood samples (approximately 0.25 mL) were then collected from the retro-orbital plexus into heparinized tubes at 0.0, 0.25, 0.5, 0.75, 1.0, 1.25, 2.0, 3.0, 5.0, 7.0, 9.0, and 11.0 h post-dosing. Plasma samples were obtained by centrifuging the blood at 4250× g for 5 min and stored frozen at −80 °C until analysis. The pharmacokinetic parameters C_{max}, t_{max}, AUC, $t_{\frac{1}{2}}$, and K_{el} were calculated using WinNonlin software.

4. Conclusions

A new rapid, sensitive, selective, reproducible, precise, and accurate UPLC-MS/MS method was developed and validated for the baricitinib plasma quantitation of. The method met the acceptance criteria for bioanalytical method validation defined by the recent FDA and EMA guidelines. The method was successfully applied in pharmacokinetic study following oral administration of a single dose of baricitinib in male rats under fasting conditions. This method demonstrates its applicability in relevant preclinical, therapeutic drug monitoring and potentially for pharmacokinetic studies of the baricitinib-drug interaction.

Author Contributions: Conceptualization, E.E. and T.E.-N.; Data curation, E.E., M.I., P.A. and T.E.-N.; Formal analysis, E.E. and M.I.; Funding acquisition, Y.A.A.; Investigation, E.E. and T.E.-N.; Methodology, E.E., M.I. and T.E.-N.; Project administration, E.E. and Y.A.A.; Software, T.E.-N.; Supervision, Y.A.A.; Validation, E.E. and M.I.; Writing—review and editing, E.E., T.E.-N., P.A. and A.A.A. All authors have read and agree to the published version of the manuscript.

Funding: This research was funded by the Research Supporting Project at King Saud University via grant number RSP/2019/45 and article processing charge (APC) was also supported by the Research Supporting Project (RSP/2019/45).

Acknowledgments: The authors extend their sincere appreciation to the Research Supporting Project at King Saud University for funding this work through the grant number RSP/2019/45.

Conflicts of Interest: The authors declare no conflict of interest.

References

1. Dugowson, C.E. 53—Rheumatoid Arthritis. In *Women and Health*; Goldman, M.B., Hatch, M.C., Eds.; Academic Press: San Diego, CA, USA, 2000; pp. 674–685. [CrossRef]
2. Zlatanovic, G.; Veselinovic, D.; Cekic, S.; Zivkovic, M.; Dordevic-Jocic, J.; Zlatanovic, M. Ocular manifestation of rheumatoid arthritis-different forms and frequency. *Bosn. J. Basic Med. Sci.* **2010**, *10*, 323–327. [CrossRef]
3. Ljung, L.; Ueda, P.; Liao, K.P.; Greenberg, J.D.; Etzel, C.J.; Solomon, D.H.; Askling, J. Performance of the Expanded Cardiovascular Risk Prediction Score for Rheumatoid Arthritis in a geographically distant National Register-based cohort: An external validation. *RMD Open* **2018**, *4*, e000771. [CrossRef] [PubMed]
4. Bartoloni, E.; Alunno, A.; Valentini, V.; Luccioli, F.; Valentini, E.; La Paglia, G.M.C.; Leone, M.C.; Cafaro, G.; Marcucci, E.; Gerli, R. Targeting Inflammation to Prevent Cardiovascular Disease in Chronic Rheumatic Diseases: Myth or Reality? *Front. Cardiovasc. Med.* **2018**, *5*, 177. [CrossRef] [PubMed]
5. Shaw, M.; Collins, B.F.; Ho, L.A.; Raghu, G. Rheumatoid arthritis-associated lung disease. *Eur. Respir Rev.* **2015**, *24*, 1–16. [CrossRef] [PubMed]
6. Murakami, K.; Kobayashi, Y.; Uehara, S.; Suzuki, T.; Koide, M.; Yamashita, T.; Nakamura, M.; Takahashi, N.; Kato, H.; Udagawa, N.; et al. A Jak1/2 inhibitor, baricitinib, inhibits osteoclastogenesis by suppressing RANKL expression in osteoblasts in vitro. *PLoS ONE* **2017**, *14*, e0181126. [CrossRef] [PubMed]
7. Mogul, A.; Corsi, K.; McAuliffe, L. Baricitinib: The Second FDA-Approved JAK Inhibitor for the Treatment of Rheumatoid Arthritis. *Ann. Pharmacother.* **2019**, *53*, 947–953. [CrossRef]
8. Zhang, X.; Chua, L.; Ernest, C., 2nd; Macias, W.; Rooney, T.; Tham, L.S. Dose/Exposure-Response Modeling to Support Dosing Recommendation for Phase III Development of Baricitinib in Patients with Rheumatoid Arthritis. *CPT Pharmacomet. Syst. Pharmacol.* **2017**, *6*, 804–813. [CrossRef]
9. Schlueter, M.; Finn, E.; Díaz, S.; Dilla, T.; Inciarte-Mundo, J.; Fakhouri, W. Cost-effectiveness analysis of baricitinib versus adalimumab for the treatment of moderate-to-severe rheumatoid arthritis in Spain. *Clin. Outcomes Res.* **2019**, *11*, 395–403. [CrossRef]
10. Wallace, D.J.; Furie, R.A.; Tanaka, Y.; Kalunian, K.C.; Mosca, M.; Petri, M.A.; Dörner, T.; Cardiel, M.H.; Bruce, I.N.; Gomez, E.; et al. Baricitinib for systemic lupus erythematosus: A double-blind, randomised, placebo-controlled, phase 2 trial. *Lancet* **2018**, *392*, 222–231. [CrossRef]
11. Shi, J.G.; Chen, X.; Lee, F.; Emm, T.; Scherle, P.A.; Lo, Y.; Punwani, N.; Williams, W.V.; Yeleswaram, S. The pharmacokinetics, pharmacodynamics, and safety of baricitinib, an oral JAK 1/2 inhibitor, in healthy volunteers. *J. Clin. Pharmacol.* **2014**, *54*, 1354–1361. [CrossRef]

12. Posada, M.M.; Cannady, E.A.; Payne, C.D.; Zhang, X.; Bacon, J.A.; Pak, Y.A.; Higgins, J.W.; Shahri, N.; Hall, S.D.; Hillgren, K.M. Prediction of Transporter-Mediated Drug-Drug Interactions for Baricitinib. *Clin. Transl. Sci.* **2017**, *10*, 509–519. [CrossRef]
13. Payne, C.; Zhang, X.; Shahri, N.; Williams, W.; Cannady, E. AB0492 Evaluation of Potential Drug-Drug Interactions with Baricitinib. *Ann. Rheum. Dis.* **2015**, *74*, 1063. [CrossRef]
14. Rodriguez-Carrio, J.; Lopez, P.; Suarez, A. Type I IFNs as biomarkers in rheumatoid arthritis: Towards disease profiling and personalized medicine. *Clin. Sci.* **2015**, *128*, 449–464. [CrossRef] [PubMed]
15. Takeuchi, T.; Genovese, M.C.; Haraoui, B.; Li, Z.; Xie, L.; Klar, R.; Pinto-Correia, A.; Otawa, S.; Lopez-Romero, P.; de la Torre, I.; et al. Dose reduction of baricitinib in patients with rheumatoid arthritis achieving sustained disease control: Results of a prospective study. *Ann. Rheum. Dis.* **2019**, *78*, 171–178. [CrossRef] [PubMed]
16. Tanaka, Y.; Atsumi, T.; Amano, K.; Harigai, M.; Ishii, T.; Kawaguchi, O.; Rooney, T.P.; Akashi, N.; Takeuchi, T. Efficacy and safety of baricitinib in Japanese patients with rheumatoid arthritis: Subgroup analyses of four multinational phase 3 randomized trials. *Mod. Rheumatol.* **2018**, *28*, 583–591. [CrossRef] [PubMed]
17. Qiu, C.; Zhao, X.; She, L.; Shi, Z.; Deng, Z.; Tan, L.; Tu, X.; Jiang, S.; Tang, B. Baricitinib induces LDL-C and HDL-C increases in rheumatoid arthritis: A meta-analysis of randomized controlled trials. *Lipids Health Dis.* **2019**, *18*, 54. [CrossRef]
18. Smolen, J.S.; Genovese, M.C.; Takeuchi, T.; Hyslop, D.L.; Macias, W.L.; Rooney, T.; Chen, L.; Dickson, C.L.; Riddle Camp, J.; Cardillo, T.E.; et al. Safety Profile of Baricitinib in Patients with Active Rheumatoid Arthritis with over 2 Years Median Time in Treatment. *J. Rheumatol.* **2019**, *46*, 7–18. [CrossRef]
19. Kim, H.; Brooks, K.M.; Tang, C.C.; Wakim, P.; Blake, M.; Brooks, S.R.; Montealegre Sanchez, G.A.; de Jesus, A.A.; Huang, Y.; Tsai, W.L.; et al. Pharmacokinetics, Pharmacodynamics, and Proposed Dosing of the Oral JAK1 and JAK2 Inhibitor Baricitinib in Pediatric and Young Adult CANDLE and SAVI Patients. *Clin. Pharmacol. Ther.* **2018**, *104*, 364–373. [CrossRef] [PubMed]
20. Adam, S.; Simon, N.; Steffen, U.; Andes, F.; Müller, D.; Culemann, S.; Andreev, D.; Hahn, M.; Scholtysek, C.; Schett, G.; et al. P148 JAK-inhibition by baricitinib and tofacitinib ameliorates pathological bone loss. *Ann. Rheum. Dis.* **2019**, *78*, 3.
21. Richez, C.; Truchetet, M.E.; Kostine, M.; Schaeverbeke, T.; Bannwarth, B. Efficacy of baricitinib in the treatment of rheumatoid arthritis. *Expert Opin. Pharmacother.* **2017**, *18*, 1399–1407. [CrossRef] [PubMed]
22. Veeraraghavan, S.; Thappali, S.R.; Viswanadha, S.; Vakkalanka, S.; Rangaswamy, M. Simultaneous Quantification of Baricitinib and Methotrexate in Rat Plasma by LC-MS/MS: Application to a Pharmacokinetic Study. *Sci. Pharm.* **2016**, *84*, 347–359. [CrossRef] [PubMed]
23. Koller, D.; Vaitsekhovich, V.; Mba, C.; Steegmann, J.L.; Zubiaur, P.; Abad-Santos, F.; Wojnicz, A. Effective quantification of 11 tyrosine kinase inhibitors and caffeine in human plasma by validated LC-MS/MS method with potent phospholipids clean-up procedure. Application to therapeutic drug monitoring. *Talanta* **2020**, *208*, 120450. [CrossRef] [PubMed]
24. Hao, Z.; Xiao, B.; Weng, N. Impact of column temperature and mobile phase components on selectivity of hydrophilic interaction chromatography (HILIC). *J. Sep. Sci.* **2008**, *31*, 1449–1464. [CrossRef] [PubMed]
25. Buszewski, B.; Noga, S. Hydrophilic interaction liquid chromatography and per aqueous liquid chromatography in fungicides analysis. *J. AOAC Int.* **2012**, *95*, 1362–1370.
26. European Medicines Agency. Olumiant, INN-Baricitinib, Summary of Product, Characteristics, 1–30. Available online: https://www.ema.europa.eu/en/documents/product-information/olumiant-epar-product-information_en.pdf (accessed on 3 February 2020).
27. Kong, R. 17—LC/MS Application in High-Throughput ADME Screen. In *Separation Science and Technology*; Ahuja, S., Dong, M.W., Eds.; Academic Press: London, UK, 2005; Volume 6, pp. 413–446.
28. Wells, D.A. Bioanalytical Applications: Solid-Phase Extraction. In *Reference Module in Chemistry, Molecular Sciences and Chemical Engineering*; Elsevier: London, UK, 2013.

29. *Guidance for Industry on Bioanalytical Method Validation*; Center for Drug Evaluation and Research, US Food and Drug Administration: Rockville, MD, USA, 2018. Available online: https://www.fda.gov/files/drugs/published/Bioanalytical-Method-Validation-Guidance-for-Industry.pdf (accessed on 2 February 2020).
30. European Medicines Agency. Guideline on Bioanalytical Method Validation. 2012. Available online: https://www.ema.europa.eu/documents/scientific-guideline/guideline-bioanalytical-method-validation_en.pdf (accessed on 5 February 2020).

Sample Availability: Samples of the compounds are not available from the authors.

© 2020 by the authors. Licensee MDPI, Basel, Switzerland. This article is an open access article distributed under the terms and conditions of the Creative Commons Attribution (CC BY) license (http://creativecommons.org/licenses/by/4.0/).

Article

Development and Validation of a Sensitive UHPLC-MS/MS Method for the Measurement of Gardneramine in Rat Plasma and Tissues and Its Application to Pharmacokinetics and Tissue Distribution Study

Nan Zhao, Hao-ran Tan, Qi-li Chen, Qi Sun, Lin Wang, Yang Song, Kamara Mohamed Olounfeh and Fan-hao Meng *

School of Pharmacy, China Medical University, Shenyang 110122, China; Zhaonan-dd@163.com (N.Z.); thr9975@163.com (H.-r.T.); chenqili000@yeah.net (Q.-l.C.); jilindaxuesunqi@126.com (Q.S.); wanglin@cmu.edu.cn (L.W.); Songyanglhyb1998@163.com (Y.S.); Mohamedkamara6994@yahoo.com (K.M.O.)
* Correspondence: fhmeng@cmu.edu.cn; Tel.: +86-24-31939448

Academic Editors: Constantinos K. Zacharis and Aikaterini Markopoulou
Received: 8 September 2019; Accepted: 30 October 2019; Published: 31 October 2019

Abstract: As a novel monoterpenoid indole alkaloid, gardneramine has been confirmed to possess excellent nervous depressive effects. However, there have been no reports about the measurement of gardneramine in vitro and in vivo. The motivation of this study was to establish and validate a specific, sensitive, and robust analytical method based on UHPLC-MS/MS for quantification of gardneramine in rat plasma and various tissues after intravenous administration. The analyte was extracted from plasma and tissue samples by protein precipitation with methanol using theophylline as an internal standard (I.S.). The analytes were separated on an Agilent ZORBAX Eclipse Plus C_{18} column using a gradient elution of acetonitrile and 0.1% formic acid in water at a flow rate of 0.3 mL/min. Gardneramine and I.S. were detected and quantified using positive electrospray ionization in multiple reaction monitoring (MRM) mode with transitions of m/z 413.1→217.9 for gardneramine and m/z 181.2→124.1 for I.S. Perfect linearity range was 1–2000 ng/mL with a correlation coefficient (r^2) of ≥0.990. The lower limit of quantification (LLOQ) of 1.0 ng/mL was adequate for application to different preclinical studies. The method was successfully applied for determination of gardneramine in bio-samples.

Keywords: gardneramine; monoterpenoid indole alkaloid; LC-MS/MS; pharmacokinetics; tissue distribution

1. Introduction

Gardneramine is a monoterpenoid indole alkaloid found in the plant species *Gardneria nutans* Sieb., *Gardneria multiflora* Makino, *Gardneria ovata* Wall, and other *Gardneria* species. [1–5]. *Gardneria* belongs to the Loganiaceae family and is widely distributed in east and southeast Asia, which is from India to central Japan and Java. All species can be found in China and were richest in Yunnan province [6,7]. Most of plants in this genus as traditional Chinese medicine, such as *Gardneria anguatifolia* Wall and *Gardneria multiflora* Makino, are employed to treat kidney deficiency, enuresis, aching lumbus and knees, wind-dampBi-syndrome, and injuries from falls [8].

Monoterpenoid indole alkaloids have long been a focus of natural product research because of their unusual carbon skeletons as well as potential bioactivities [9–11]. Gardneramine as an active monoterpenoid indole alkaloid has been shown to have excellent central depressive effect [12], to monoterpenoid indole alkaloid has been shown to have excellent central depressive effect [12], to affect neuromuscular transmission [13], to produce a hypotensive effect like papaverine derived from peripheral vasodilatation, to have direct depressive action on myocardium, and to have central depressive action [14]. Previous studies indicated that gardneramine possesses a hexamethonium-like action on superior cervical ganglion and depresses the contraction of the organ through an inhibition of the parasympathetic ganglionic transmission [15,16]. Moreover, gardneramine was known to be an antagonist of nicotinic receptor by selectively inhibiting dimethylphenylpiperazinium-induced contraction [17]. These significant findings provide the basis for the further in-depth drug-targeted research of gardneramine as a novel anesthetic agent.

To our knowledge, the preclinical pharmacokinetic studies play a main role that can be used to make critical decisions supporting the safety and efficacy of the development of a new drug. In view of gardneramine as a potential anesthetic agent, it is pivotal important to understand its absorption and distribution. However, there are no report in the pharmacokinetics and tissue distribution studies of gardneramine in bio-samples. Therefore, there is a need to understand the characterization and diversity of gardneramine and it is necessary and meaningful to establish a more accurate and selective bioanalytical method for the determination of gardneramine in plasma and various tissues.

Nowadays, UHPLC-MS/MS is a powerful technique used for drug research in vivo. Therefore, the aim of this investigation was to establish and validate a sensitive and credible approach for analyzing gardneramine in rat plasma and organs by using an UHPLC-MS/MS. Applying this method, we conducted the preclinical pharmacokinetics and tissue distribution characteristics of gardneramine in rats after intravenous administration.

2. Results and Discussion

2.1. Optimization of LC-MS/MS Conditions

To optimize MS/MS parameters, the feasibility of electrospray in both positive and negative ion modes were investigated. Gardneramine and I.S. were found to show higher response in positive ion mode. MRM mode was used to select the precursor ions and product ions, which were shown in Figure 1. A precursor ion and two MRM transitions have been established for gardneramine. The MRM transitions for gardneramine and I.S. were m/z 413.1→217.9 and m/z 181.2→124.1, respectively, which were employed for quantitative analysis. The qualifier ion for gardneramine was set at m/z 232.9. The parameters, such as declustering potential (DP) and collision energy (CE), were optimized to acquire higher sensitivity, which are shown in Table 1.

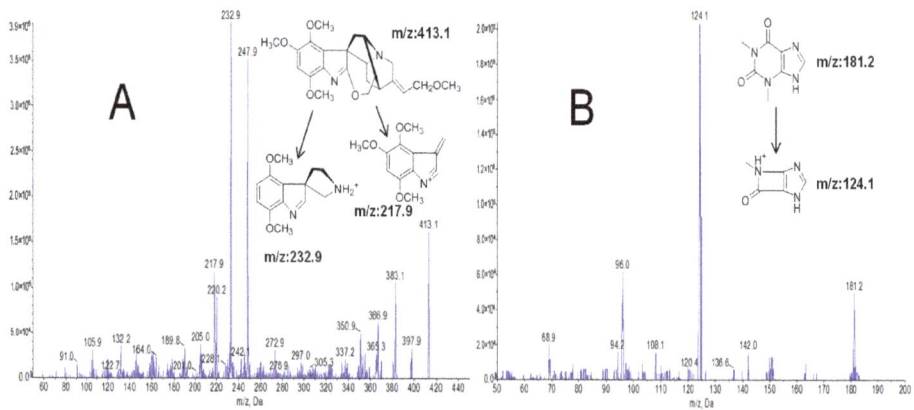

Figure 1. Mass spectra and structures of gardneramine (A) and internal standard I.S (B).

Table 1. Optimized mass spectrometry conditions for the determination of gardneramine and I.S.

Table 1. Optimized mass spectrometry conditions for the determination of gardneramine and I.S.

Analytes	Precursor Ion (m/z)	Product Ion (m/z)	Declustering Potential (V)	Collision Energy (eV)
gardneramine	413.1	217.9	134.8	36.6
		232.9	123.9	46
I.S.	181.2	124.1	70	26.6

We also attempted to improve the mobile phase system in several trials. To obtain a satisfactory chromatograph, acetonitrile and 0.1% formic acid phase were adopted as the mobile phase. The column temperature, flow rate, and run time were optimized as 30 °C and 0.3 mL/min, respectively. The mobile phase additive, column temperature, flow rate, and run time were optimized. Finally, a gradient behavior was selected, and total run time was finished in 5.5 min.

2.2. Optimization of the Extraction Method

Protein precipitation and liquid–liquid extraction was considerably compared to select a suitable method for sample pretreatment. Ethyl acetate, dichloromethane, and tert-butyl ether as extracting agents produced low recovery and detectable interference from plasma matrix. Thus, protein precipitation was selected to perform the quantitative analysis. Methanol and acetonitrile were compared to as suitable solvent for protein precipitation and deproteinization with methanol provided a high extraction recovery of more than 88.03% for gardneramine.

2.3. Method Validation

Typical MRM chromatograms of gardneramine and theophylline (I.S.) are shown in Figure 2, which were involved blank plasma, blank plasma spiked with gardneramine and I.S., and the plasma sample obtained from a rat 2 h after intravenous administration of gardneramine. Gardneramine and I.S. were eluted at retention times of 2.61 and 1.88 min, respectively. Under the validated HPLC-MS/MS conditions, no interference peaks were observed at the retention time of gardneramine and I.S. This result revealed that bio-samples could be accurately differentiated and quantified with this method.

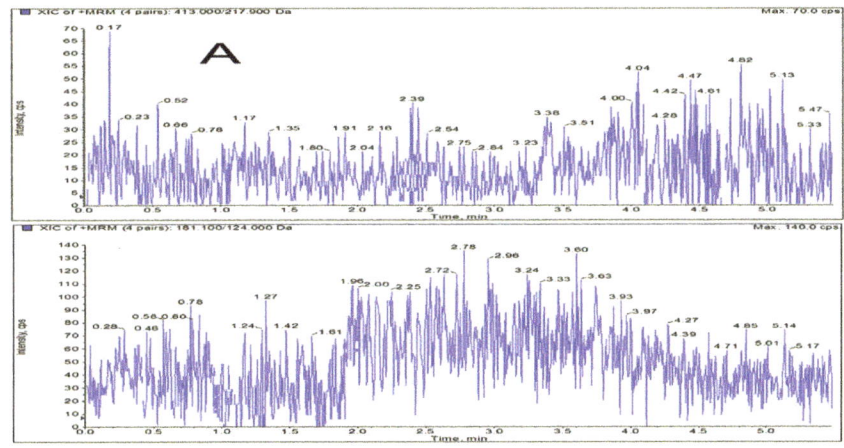

Figure 2. Cont.

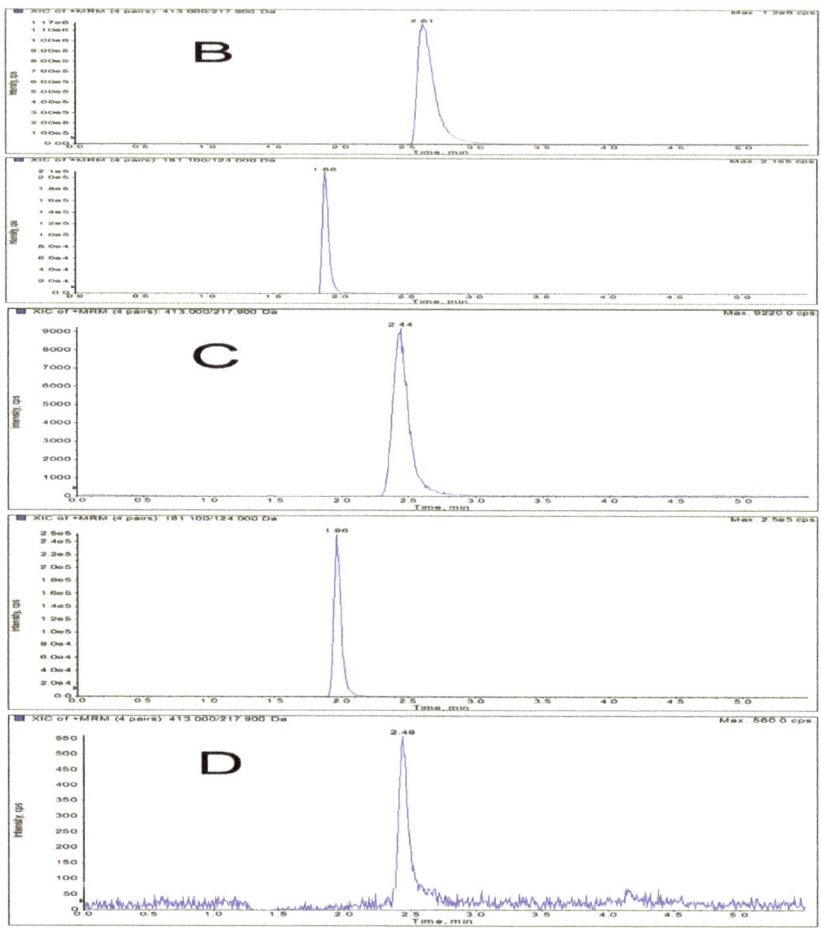

Figure 2. Typical chromatograms of gardneramine and theophylline (I.S.) in rat plasma samples: (**A**) blank rat plasma; (**B**) blank rat plasma spiked with gardneramine (2.61 min) and I.S. (1.88 min); (**C**) plasma sample from the pharmacokinetic study; and (**D**) blank rat plasma spiked with gardneramine at lower limit of quantification (LLOQ).

The calibration curves, correlation coefficients, linear ranges, and lower limit of quantification (LLOQ) of gardneramine in the rat plasma and tissues homogenates were presented in Table 2. The calibration ranging from 1 to 2000 ng/mL yielded a good linearity ($r^2 \geq 0.990$) corresponding to the range for gardneramine. The lower limit of quantification (LLOQ) for gardneramine was 1 ng/mL.

A summary of precision and accuracy of gardneramine in rat plasma and tissues is shown in Table 3. The data suggested that the precision and accuracy of the method were sufficient for the quantification of gardneramine in rat plasma and tissues.

The results of extraction recovery and matrix effects are presented in Table 4, which suggest that the endogenous matrix had no significant impact on the measurement of gardneramine in biological samples, and the recoveries of gardneramine at four levels were no less than 88%, indicating that the method of extraction we selected was dependable.

Stability data of gardneramine in rat plasma and different tissues are listed in Table 5 and show that gardneramine was stably detected in bio-samples stored under various conditions.

The results of dilution integrity experiments indicated that the accuracy of measured concentration was below 9.8 with a precision of no more than 3.8, which was acceptable in dilution integrity analysis.

Table 2. Standard curves and LLOQ of gardneramine in biological samples.

Samples	Calibration Curves	Correlation Coefficients (r)	Linear Ranges (ng/mL)	LLOQs (ng/mL)
Plasma	Y = 0.0720 + 0.18447x	0.9924	1–2000	1
Heart	Y = −0.0943 + 0.13776x	0.9931	1–2000	1
Liver	Y = 0.0836 + 0.19987x	0.9906	1–2000	1
Spleen	Y = −0.0286 + 0.20747x	0.9913	1–2000	1
Lung	Y = 0.0996 + 0.09197x	0.9954	1–2000	1
Kidney	Y = 0.0193 + 0.25430x	0.9941	1–2000	1
Brain	Y = −0.0749 + 0.14529x	0.9917	1–2000	1
Intestine	Y = 0.0974 + 0.19646x	0.9962	1–2000	1
Stomach	Y = 0.0432 + 0.37274x	0.9901	1–2000	1

Table 3. Precision and accuracy of gardneramine in rat plasma and tissue homogenates ($n = 6$).

Samples	Spiked Concentration (ng/mL)	Intraday		Interday	
		Precision (RSD, %)	Accuracy (Mean%)	Precision (RSD, %)	Accuracy (Mean%)
Plasma	1	8.94	0.98	2.31	−1.38
	3	6.09	3.13	2.30	5.49
	45	4.89	1.33	6.27	4.29
	1600	5.80	3.65	3.47	4.89
Heart	1	3.05	1.02	4.03	4.21
	3	4.78	−1.42	3.75	−2.88
	45	4.53	1.33	4.08	3.18
	1600	3.88	3.79	2.81	3.88
Liver	1	3.64	1.02	3.27	3.62
	3	4.58	2.42	3.12	2.28
	45	2.57	0.33	2.13	0.17
	1600	3.96	0.53	5.07	−1.61
Spleen	1	3.34	0.33	7.10	−2.91
	3	3.07	−0.78	4.47	1.23
	45	2.24	−0.28	4.44	1.40
	1600	3.50	−1.39	1.04	−1.66
Lung	1	4.67	1.02	1.03	2.81
	3	3.56	2.84	2.24	−5.76
	45	1.69	−1.00	4.15	−2.26
	1600	3.50	1.84	3.14	2.87
Kidney	1	2.14	0.88	2.27	1.43
	3	2.74	3.55	4.96	5.53
	45	3.51	0.76	5.93	−1.72
	1600	9.59	2.32	4.38	4.77
Brain	1	4.47	1.03	2.64	3.96
	3	2.58	3.12	6.32	6.61
	45	4.86	−2.06	1.93	−2.77
	1600	10.24	0.47	4.07	−1.27
Intestine	1	4.76	0.92	7.77	−5.36
	3	3.26	−0.03	12.32	−4.68
	45	6.20	0.02	5.09	0.84
	1600	3.32	4.66	2.98	6.13
Stomach	1	4.55	0.16	3.50	−1.08
	3	5.27	3.20	2.40	8.17
	45	3.58	0.20	8.53	−2.49
	1600	11.08	−6.79	4.05	−8.04

Table 4. Matrix effect and extraction recovery of gardneramine in rat plasma and tissue homogenates ($n = 5$).

Samples	Spiked Concentration (ng/mL)	Matrix Effect		Extraction Recovery	
		Mean ± SD (%)	RSD (%)	Mean ± SD (%)	RSD (%)
Plasma	1	101.1 ± 5.8	5.8	101.6 ± 6.6	6.5
	3	101.4 ± 7.1	7.0	99.0 ± 7.5	7.5
	45	103.1 ± 6.9	6.7	93.5 ± 6.2	6.6
	1600	99.5 ± 4.1	4.1	99.5 ± 3.3	3.3
Heart	1	92.1 ± 3.5	3.8	94.0 ± 8.1	8.7
	3	99.8 ± 5.2	5.2	101.3 ± 5.9	5.8
	45	95.8 ± 6.1	6.4	99.9 ± 6.9	6.9
	1600	106.5 ± 9.6	9.0	95.3 ± 1.1	1.1
Liver	1	100.6 ± 1.3	1.3	99.5 ± 1.4	1.4
	3	102.0 ± 4.9	4.8	104.3 ± 3.7	3.5
	45	102.9 ± 1.9	1.9	99.6 ± 2.6	2.6
	1600	95.9 ± 3.4	3.5	102.6 ± 5.6	5.4
Spleen	1	96.3 ± 2.1	2.2	99.1 ± 2.3	2.4
	3	84.0 ± 3.5	4.2	111.1 ± 6.1	5.5
	45	113.05 ± 8.5	7.5	100.8 ± 9.9	9.9
	1600	111.4 ± 6.5	5.8	102.9 ± 4.4	4.2
Lung	1	99.9 ± 0.8	0.8	98.8 ± 5.0	5.1
	3	104.3 ± 2.7	2.6	103.1 ± 9.4	9.2
	45	101.2 ± 3.8	3.8	100.6 ± 3.9	3.8
	1600	100.8 ± 0.5	0.5	102.3 ± 3.7	3.6
Kidney	1	97.9 ± 3.6	3.6	101.5 ± 3.1	3.0
	3	95.6 ± 5.4	5.6	96.9 ± 5.4	5.6
	45	104.5 ± 3.9	3.7	99.4 ± 6.1	6.1
	1600	111.3 ± 2.0	1.8	91.6 ± 3.0	3.2
Brain	1	103.0 ± 9.1	8.8	95.0 ± 10.7	11.3
	3	106.2 ± 3.3	3.1	100.7 ± 5.9	5.9
	45	93.9 ± 3.5	3.8	103.3 ± 2.6	2.5
	1600	105.2 ± 1.0	1.0	100.4 ± 1.7	1.7
Intestine	1	100.8 ± 3.7	3.7	101.8 ± 5.5	5.4
	3	98.7 ± 1.5	1.5	88.0 ± 2.0	2.3
	45	100.3 ± 5.9	5.9	98.9 ± 4.0	4.0
	1600	96.0 ± 5.3	5.5	93.0 ± 4.5	4.8
Stomach	1	103.5 ± 3.4	3.3	100.0 ± 4.0	4.0
	3	94.6 ± 4.0	4.2	107.3 ± 1.6	1.5
	45	107.4 ± 5.9	5.5	105.9 ± 4.6	4.4
	1600	99.2 ± 5.2	5.3	105.3 ± 5.5	5.3

Table 5. Stability test of gardneramine in rat plasma and tissue homogenates ($n = 5$).

Samples	Spiked CONC (ng/mL)	Short-Term (at Room Temperature for 4 h)		Autosampler 4 °C for 24 h		Three Freeze-Thaw Cycles		Long-Term (at −20 °C for 30 days)	
		RE [1] (%)	RSD (%)	RE (%)	RSD (%)	RE (%)	RSD (%)	RE (%)	RSD (%)
Plasma	3	−10.2	6.62	−0.37	8.27	−3.45	5.06	12.8	4.55
	45	−4.78	1.50	−2.49	5.50	−1.72	5.64	1.37	4.01
	1600	−1.15	2.37	−1.56	3.69	−4.36	2.18	−1.11	7.07
Heart	3	4.33	5.94	2.20	2.11	3.59	3.60	0.11	3.81
	45	−11.7	2.60	−10.7	6.66	−7.75	4.44	4.49	3.37
	1600	0.77	3.32	−4.75	6.08	−4.89	2.25	−1.86	5.10
Liver	3	−3.00	6.03	4.20	6.56	3.57	3.26	−2.75	6.10
	45	1.76	3.19	0.66	3.75	5.36	6.22	−5.4	2.31
	1600	−2.10	2.99	2.09	2.32	−0.10	3.96	4.12	5.68
Spleen	3	−0.92	2.90	−4.67	5.49	−1.69	5.40	−7.62	2.38
	45	8.75	2.16	5.20	3.68	11.4	3.29	12.5	3.37
	1600	12.2	2.97	4.66	3.59	9.10	4.58	3.67	1.77
Lung	3	7.71	5.70	7.86	3.79	9.81	3.23	3.03	1.13
	45	0.94	7.00	2.15	5.19	6.06	1.60	−2.63	2.31
	1600	0.47	1.49	4.32	4.12	−1.53	2.97	0.59	5.68
Kidney	3	2.77	3.63	−8.30	2.61	−3.34	6.06	−8.27	2.52
	45	6.29	1.18	−4.53	8.27	−6.90	3.19	−8.88	1.96
	1600	−0.45	2.08	−10.36	4.59	6.73	4.85	0.49	3.49
Brain	3	0.49	4.66	4.25	7.41	−4.07	2.50	−7.66	2.52
	45	3.50	4.19	4.25	2.00	4.02	5.85	4.91	5.13
	1600	1.20	5.52	−1.89	6.80	3.52	1.85	9.09	4.41
Intestine	3	−4.88	4.45	−1.67	8.33	0.97	5.99	6.04	3.51
	45	−2.93	4.54	−6.52	4.54	−5.83	4.03	−9.83	5.02
	1600	−2.07	3.71	4.35	1.49	−2.42	5.78	−0.41	2.68
Stomach	3	−3.04	2.28	−3.46	1.76	−1.04	4.59	−0.18	6.49
	45	3.71	2.75	1.39	4.53	−2.04	4.88	6.23	2.24
	1600	−11.7	4.26	0.55	6.57	−11.3	6.76	−5.87	7.83

[1] RE was relative error.

2.4. Pharmacokinetic Study

The validated UHPLC-MS/MS approach was successfully applied to pharmacokinetic study of gardneramine after intravenous administration of 10 mg/kg to rats. The mean plasma concentration-time profiles of gardneramine and major pharmacokinetic parameters are depicted in Figure 3. The pharmacokinetic parameters were estimated by using non-compartment model and are summarized in Table 6. After the intravenous administration of gardneramine in rats, elimination was fast with $t_{1/2}$ of 1.94 h, indicating that the residence time of gardneramine was short. Moreover, rapid decline of the plasma concentration could be partly due to tissue distribution quickly, which was demonstrated by V_d of 19.30 ± 2.88 L/kg. V_d reflected the tissue uptake after intravenous administration, and the higher the value, the wider the distribution. At present, this is the first study about pharmacokinetics characteristics of gardneramine in vivo, and the observations suggested that the validated analytical UHPLC-MS/MS method was suitable and sufficient.

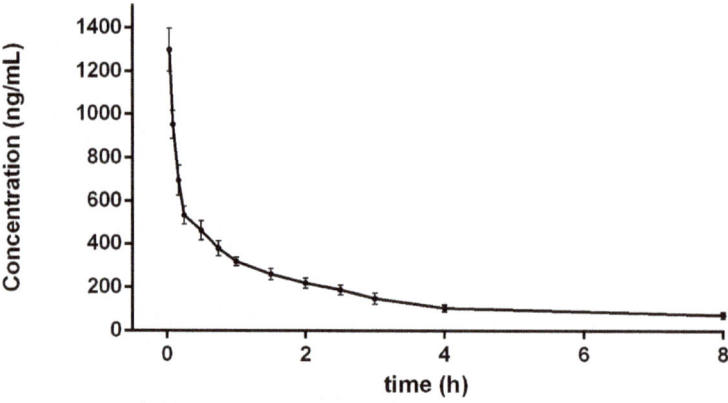

Figure 3. Plasma concentration-time profiles of gardneramine after intravenous injection of gardneramine at dose of 10 mg/kg (n = 12).

Table 6. The pharmacokinetic parameters of gardneramine in rat plasma (n = 12).

Parameters	Value
$t_{1/2}$ (h)	1.94 ± 0.46
AUC_{0-t} (ng h/mL)	843.36 ± 54.02
$AUC_{0-\infty}$ (ng h/mL)	1443.88 ± 207.96
MRT_{iv} (h)	2.70 ± 0.64
CL (L/h/kg)	7.07 ± 1.00
V_d (L/kg)	19.30 ± 2.88

2.5. Tissue Distribution

The established LC-MS/MS method was successfully applied for determination of gardneramine in rat tissues after an intravenous administration of 10 mg/kg gardneramine, as depicted in Figure 4. After intravenous injection, gardneramine underwent a wide distribution into all tested tissues, especially mainly into the intestine. The highest concentration of gardneramine was observed in the intestine (40.7 ± 3.2 µg/g, 0.5 h), followed consequently by intestine (22.3 ± 2.8 µg/g, 1 h), intestine (16.7 ± 1.8 µg/g, 2 h), lung (15.8 ± 1.2 µg/g, 1 h), and liver (13.1 ± 0.9 µg/g, 0.5 h), which might have been attributed to the high blood flow in these organs. We speculate that these organs may be the targets of gardneramine and hypothesize a first rapid passage of gardneramine from the blood to the intestine and to the stomach partially, from which the substance would then be reabsorbed to reach the lung and kidney, where the concentration reaches a maximum after 1 h from administration. As we know, gut microbiota communicates with the central nervous system [18–21]. Gardneramine has a good ability to be absorbed in the intestine, suggesting that it perhaps influences the brain function and behavior by the gut-brain axis, which requires further researches. Meanwhile, the concentration of gardneramine in brain was increased within 2 h (4.5 ± 0.2 µg/g, 2 h) and then decreased in the following 2 h, which indicated that it could pass through the blood-brain barrier and did not accumulate during the detection period. The result was suggested that gardneramine could have potential to treat brain-related diseases, which was in coincidence with its central nervous activities. However, these findings require confirmation by further researches.

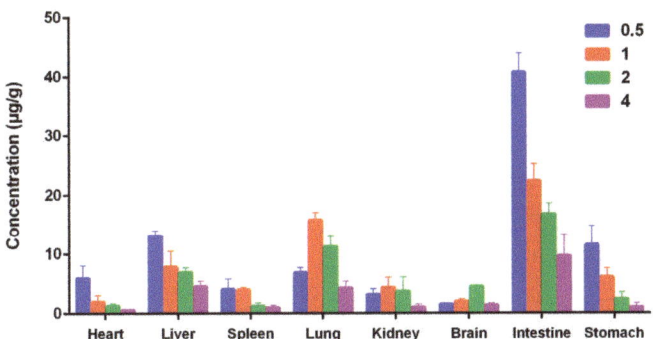

Figure 4. Tissue distribution in rats after a single intravenous administration of 10 mg/kg gardneramine (n = 5).

3. Materials and Methods

3.1. Chemicals and Reagents

Gardneramine (purity over 98%, by HPLC, Agilent Technologies, Palo Alto, CA, USA) was isolated from the roots of *Gardneria multiflora* Makino., and its structure was identified by detailed HR-MS and NMR analysis (Bruker, Karlsruhe, Germany). Theophylline as internal standard was purchased from Sigma-Aldrich (Shanghai, China). HPLC-grade acetonitrile and formic acid were purchased from Fisher Scientific (Tustin, CA, USA). Ultra-pure water, which was prepared by a Milli-Q purification system (Bedford, MA, USA), was used for the mobile phase.

3.2. Animals

Sprague Dawley (SD) rats (weighing 220 ± 20 g) were provided by the Experimental Animal Centre of China Medical University (Shenyang, China). All protocols and care of the rats were performed in compliance with the Guidelines for Care and Use of Laboratory Animals of China Medical University and authorized by the Animal Ethics Committee of the institution (code number: CMU2019202). The rats were acclimatized in an environmentally controlled breeding room in a 12 h light/dark cycle for one week at 22 ± 2 °C and 55 ± 10% relative humidity, with food and water provided ad libitum prior to testing, and then fasted with free access to water for 12 h before drug administration.

3.3. Instrumentation and Conditions

Samples were analyzed using an Agilent series 1290 UHPLC system (Agilent Technologies, Palo Alto, CA, USA) equipped with an AB 3500 triple-quadrupole mass spectrometer (AB Sciex, Ontario, Canada) and an electrospray source. The separation was carried out by using an Agilent ZORBAX Eclipse Plus C$_{18}$ column (100 mm × 2.1 mm i.d., 1.8 μm) and the gradient elution program was as follows: 20–20% A (0–0.2 min), 20–95% A (0.2–2.3 min), 95–95% A (2.3–4.4 min), 20–20% A (4.5–5.5 min). The flow rate was 0.3 mL/min and injection volume was 5.0 μL. Mobile phase consists of acetonitrile (A) and 0.1% formic acid aqueous solution (B). The mass spectrometer (AB Sciex, Concord, ON, Canada) was operated under multiple reaction monitoring (MRM) in a positive ionization mode using the following conditions: ion spray voltage of 5.5 kV, turbo spray temperature of 530 °C, gas 1 at 35.0 psi, gas 2 at 30.0 psi, curtain gas at 35.0 psi, and collision gas at 8.0 psi. The transitions (precursor-product) monitored were at m/z 413.1→217.9 and 413.1→232.9 for gardneramine and at m/z 181.2→124.1 for theophylline (I.S.). The optimized precursor and product ion-pairs and the mass spectrometer parameters are summarized in Table 1.

3.4. Preparation of Calibration Standards and Quality Control Samples

The stock solutions of gardneramine and I.S. were prepared by dissolving each reagent in methanol with the concentration of 1.00 mg/mL and stored at −25 °C. Calibration working solutions were achieved at 1.0, 5.0, 10.0, 40.0, 160.0, 400, 800, and 2000 ng/mL by diluting the stock solutions with methanol. The quality control samples were prepared at concentrations of 3.0, 45.0, and 1600 ng/mL for plasma and tissue samples. The intermediate stock solutions and working solutions were stored at 4 °C until use for no longer than 4 weeks.

3.5. Sample Preparation

The 100 µL of plasma sample, 50 µL of I.S., and 200 µL of methanol were added in 500 µL microcentrifuge tubes and then extracted by vortexing for 1 min to precipitate proteins. After a 10 min centrifugation step at 12,000× g, the upper organic layer was transferred to another tube and evaporated to dryness with a stream of nitrogen at 35 °C. The residues were reconstituted in 100 µL of initial mobile phase and centrifuged at 12,000× g for 10 min; 10 µL of the supernatant was transferred to sampling vials for the LC-MS/MS system.

The tissue samples (1.0 g) were homogenized in normal saline (5 mL). Homogenized tissues were centrifuged at 12,000× g for 10 min, and then, the upper layer was transferred to another tube. Subsequently, 100 µL of supernatant was treated in the same manner as the plasma sample.

3.6. Method Validation

Based on the FDA and EMA guidelines for bioanalytical method validation, the indicators of evaluation including selectivity, linearity, precision, accuracy, extraction recovery, matrix effect, and stability were followed for all experiments.

3.6.1. Selectivity

The selectivity was assessed by comparing the chromatograms of blank rat plasma from six different sources, blank plasma-spiked gardneramine and theophylline, and actual samples after administration of gardneramine to investigate the potential interferences at the retention times of gardneramine and I.S.

3.6.2. Linearity and Sensitivity

The calibration curves were generated by analyzing spiked calibration samples, ranging from 1 ng/mL to 2 µg/mL. The linearity of calibration curves was assessed by plotting the peak area ratio of gardneramine to I.S. versus the nominal concentration of gardneramine by weighted ($1/x^2$) least square linear regression. The lower limit of quantification (LLOQ) was defined as the lowest concentration on the standard curve with a signal-to-noise (S/N) ratio > 5 and both the precision and accuracy <20%.

3.6.3. Precision and Accuracy

The intra-day and inter-day precision and accuracy were evaluated by analyzing QC samples and LLOQ concentrations on three consecutive days ($n = 6$). Precision was calculated as relative standard deviation (RSD), while the accuracy was measured as relative error (RE).

3.6.4. Extraction Recovery and Matrix Effects

For extraction recovery and matrix effect analysis, QC samples with three concentrations and LLOQ level were applied. Recovery was evaluated by comparing the responses of gardneramine from QC samples pre-spiked in blank sample with those of post-extracted blank plasma spiked at the same concentration ($n = 5$). The matrix effect was quantified by the comparison of the peak areas of post-extracted spiked rat plasma with those of equivalent concentrations of pure standard solutions ($n = 5$).

3.6.5. Stability

Stability test of gardneramine was evaluated by examining three concentrations of QC samples with five replicates under different storage conditions: (1) freeze-thaw cycles stability (three cycles), (2) at room temperature for 4 h, (3) at 4 °C for 24 h, and (4) at −80 °C for 4 weeks.

3.6.6. Dilution Integrity

The concentrations of gardneramine in some tissue samples are higher than the upper limit of quantitation (ULOQ). Experiments was performed by a 4.5-fold dilution of ULOQ concentration ($n = 6$), with both of accuracy and precision no more than 15%.

3.7. Pharmacokinetic Study

Gardneramine was dissolved in polyethylene glycol (PEG)-saline (1:4, PEG: 380–420) for the preparation of dosing solutions. Twelve SD rats (220 ± 20 g) were intravenously administered with a single dose of 10 mg/kg. Blood samples (about 0.25 mL) were collected from suborbital vein of each rat at 2, 5, 10, 15, 30, 45, 60, 90, 120, 150, 180, 240, and 480 min after intravenous administration. The samples were placed in heparinized centrifuge tubes and immediately separated by centrifugation at 12,000× g for 10 min, and then, the supernatant fractions were transferred and stored at −20 °C until further analysis.

3.8. Tissue Distribution Study

Twenty rats were randomly divided into four groups ($n = 5$ per group), and intravenous dose was 10 mg/kg. Different tissues were harvested at each time point (30, 60, 120, and 240 min after administration) and prepared by the methods described in Section 3.5: Sample Preparation.

4. Conclusions

An accurate, sensitive, and efficient UHPLC-MS/MS method was established and validated for the measurement of gardneramine in bio-samples for the first time, and it was successfully used to evaluate pharmacokinetics and tissue distribution profile of gardneramine after intravenous administration to rats. The results suggested that gardneramine was well absorbed in plasma and all the other tissues tested in this study. Additionally, gardneramine can be easily taken up by intestine and brain, which revealed that gardneramine may be useful to the development of brain-related diseases therapeutics. Therefore, these findings will help to provide a reference for safe and effective drug doses for the further pharmacological and clinical practice of gardneramine and structural analogues.

Author Contributions: Conceptualization, N.Z. and F.M.; methodology, Q.S. and Q.-l.C.; software, L.W.; validation, N.Z. and K.M.O.; formal analysis, Y.S. and N.Z.; investigation, H.-r.T.; writing—original draft preparation, N.Z., H.-r.T., and K.M.O.; writing—review and editing, F.M.; funding acquisition, N.Z. and F.-h.M.

Funding: This research was funded by the National Natural Science Foundation of China, grant numbers 31600278 and 81573687 and by the Natural Science Foundation of Liaoning Province, China, grant number 201602880.

Acknowledgments: This work was supported by the National Natural Science Foundation of China (No. 31600278 and 81573687) and by the Natural Science Foundation of Liaoning Province, China (No. 201602880).

Conflicts of Interest: The authors declare no conflict of interest.

References

1. Haginiwa, J.; Sakai, S.; Kubo, A.; Hamamoto, T. Alkaloids of Gardneria nutans sieb. et Zucc. *Yakugaku Zasshi* **1967**, *87*, 1484–1488. [CrossRef]
2. Sakai, S.; Aimi, N.; Yamaguchi, K.; Ohhira, H.; Hori, K.; Haginiwa, J. Gardneria alkaloids-IX structures of chitosenine and three other minor bases from Gardneria multiflora Makino. *Tetrahedron Lett.* **1975**, *16*, 715–718. [CrossRef]

3. Li, X.N.; Cai, X.H.; Feng, T.; Li, Y.; Liu, Y.P.; Luo, X.D. Monoterpenoid indole alkaloids from Gardneria ovate. *J. Nat. Prod.* **2011**, *74*, 1073–1078. [CrossRef]
4. Yang, W.X.; Chen, Y.F.; Yang, J.; Huang, T.; Wu, L.L.; Xiao, N.; Hao, X.J.; Zhang, Y.H. Monterpenoid indole alkaloids from Gardneria multiflora. *Fitoterapia* **2018**, *124*, 8–11. [CrossRef] [PubMed]
5. Haginiwa, J.; Sakai, S.; Kubo, A.; Takahashi, K.; Taguchi, M. Gardneria alkaloids-IV comparative study of alkaloids on Gardneria nutans sieb. et Zucc., G. multiflora Makino, G. shimadai hayata and so-called G. insularis nakai. *Yakugaku Zasshi* **1970**, *90*, 219–223. [CrossRef] [PubMed]
6. Chen, X.M. The geographical distribution of the loganiaceae in China. *J. South China Agric. Univ.* **1995**, *16*, 92–97.
7. Editorial Committee of Flora of China Chinese Academy of Sciences. *Flora of China*; Science Press: Beijing, China, 1999; Volume 61, p. 242.
8. Wang, G.Q. *National Compilation of Chinese Herbal Medicine*, 3rd ed.; People's Medical Publishing House: Beijing, China, 2014; Volume 3, p. 616.
9. Frederich, M.; Jacquier, M.; Thepenier, P.; Mol, P.D.; Tits, M.; Genevieve, P.; Clement, D.; Luc, A.; Monique, Z. Antiplasmodial activity of alkaloids from various strychnos species. *J. Nat. Prod.* **2002**, *65*, 1381–1386. [CrossRef] [PubMed]
10. Sarah, E.O.C.; Justin, J.M. Chemistry and biology of monoterpene indole alkaloid biosynthesis. *Nat. Prod. Rep.* **2006**, *23*, 532–547.
11. Passos, C.D.S.; Soldi, T.C.; Abib, R.T.; Apel, M.A.; Claudia, S.P.; Marcourt, L.; Gottfried, C.; Henriques, A.T. Monoamine oxidase inhibition by monoterpene indole alkaloids and fractions obtained from Psychotria suterella and Psychotria laciniata. *J. Enzyme Inhib. Med. Chem.* **2013**, *28*, 611–618. [CrossRef] [PubMed]
12. Harada, M.; Ozaki, Y.; Murayama, S.; Sakai, S.; Haginiwa, J. Pharmacological studies on Gardneria alkaloids.I. central effects. *Yakugaku Zasshi* **1971**, *91*, 997–1003. [CrossRef] [PubMed]
13. Harada, M.; Ozaki, Y. Effect of indole alkaloids from Gardneria genus and Uncaria genus on neuromuscular transmission in the rat limb in situ. *Chem. Pharm. Bull.* **1976**, *24*, 211–214. [CrossRef] [PubMed]
14. Harada, M.; Ozaki, Y. Pharmacological studies on Gardneria alkaloids.II. peripheral effects (effects on circulatory and digestive systems). *Yakugaku Zasshi* **1972**, *92*, 1540–1546. [CrossRef] [PubMed]
15. Harada, M.; Ozaki, Y. Effect of Gardneria alkaloids on ganglionic transmission in the rabbit and rat superior cervical ganglia in situ. *Chem. Pharm. Bull.* **1978**, *26*, 48–52. [CrossRef] [PubMed]
16. Harada, M.; Ozaki, Y.; Ohno, H. Effects of indole alkaloids from Gardneria nutans sieb. et Zucc. and Uncaria rhynchophylla Mio. on a guinea pig urinary bladder preparation in situ. *Chem. Pharm. Bull.* **1979**, *27*, 1069–1074. [CrossRef] [PubMed]
17. Ozaki, Y.; Harada, M. Site of the ganglion blocking action of gardneramine and hirsutine in the dog urinary bladder in situ preparation. *Japan. J. Pharmacol.* **1983**, *33*, 463–471. [CrossRef] [PubMed]
18. Cryan, J.F.; Dinan, T.G. Mind-altering microorganisms: The impact of the gut microbiota on brain and behavior. *Nat. Rev. Neurosci.* **2012**, *13*, 701–712. [CrossRef] [PubMed]
19. Bravo, J.A.; Forsythe, P.; Chew, M.V.; Escaravage, E.; Savignac, H.M.; Dinan, T.G.; Bienenstock, J.; Cryan, J.F. Ingestion of Lactobacillus strain regulates emotional behavior and central GABA receptor expression in a mouse via the vagus nerve. *Proc. Natl. Acad. Sci. USA* **2011**, *108*, 16050–16055. [CrossRef] [PubMed]
20. Bercik, P.; Denou, E.; Collins, J.; Jackson, W.; Lu, J.; Jury, J.; Deng, Y.; Blennerhassett, P.; Macri, J.; Mccoy, K.D.; et al. The intestinal microbiota affect central levels of brain-derived neurotropic factor and behavior in mice. *Gastroenterology* **2011**, *141*, 599–609. [CrossRef] [PubMed]
21. Bravo, J.A.; Marcela, J.P.; Forsythe, P.; Kunze, W.; Dinan, T.G.; Bienenstock, J.; Cryan, J.F. Communication between gastrointestinal bacteria and the nervous system. *Curr. Opin. Pharmacol.* **2012**, *12*, 667–672. [CrossRef]

Sample Availability: Samples of the compounds gardneramine and theophylline are available from the authors.

© 2019 by the authors. Licensee MDPI, Basel, Switzerland. This article is an open access article distributed under the terms and conditions of the Creative Commons Attribution (CC BY) license (http://creativecommons.org/licenses/by/4.0/).

of eluent composition examined ($r \geq 0.98$). All R_M values are the average of three measurements, with relative standard deviations (RSDs) lower than 4.5%; the 95% confidence interval associated with each R_{Mw} value never exceeded 0.16 (Table 1).

Table 1. Experimental parameters obtained by chromatographic techniques.

Analyte	R_{Mw} [1]	$\log k_{w\ C18}$ [2]	$\log k_{w\ IAM}$ [3]	PPB_{HSA} [4] (%)	PPB_{AGP} [5] (%)
6-thioguanine	0.21	0.32	0.03	24.11	45.96
6-mercaptopurine	0.23	0.45	−0.19	20.08	22.15
Folic acid	0.57	0.99	−0.26	69.40	3.45
Azathioprine	1.41	1.34	0.68	49.22	75.96

[1] R_{Mw}—chromatographic parameter of hydrophobicity obtained by RP-TLC; [2] $\log k_{w\ C18}$—chromatographic parameter of hydrophobicity obtained by RP-HPLC; [3] $\log k_{w\ IAM}$—chromatographic parameter of lipophilicity obtained by IAM-HPLC; [4] PPB_{HSA}—percentage of analyte bound to human serum albumin (HSA) obtained by HSA-HPLC; [5] PPB_{AGP}—percentage of analyte bound to α1-acid glycoprotein (AGP) obtained by AGP-HPLC.

Further to the results presented above, all investigated compounds were expected to show high affinity for chromatographic column packing consisting of octadecylsilane-silica particles. The obtained retention times, t_R, were used to calculate $\log k$ values according to Equation (3):

$$\log k = \log \frac{t_R - t_0}{t_0}, \tag{3}$$

with t_0 being the retention time of an unretained solute, sodium nitrate. Furthermore, chromatographic parameter of hydrophobicity, $\log k_{w\ C18}$, was obtained by extrapolation to pure buffer using the linear Equation (4):

$$\log k = \log k_w - S\Phi, \tag{4}$$

where S is the slope of the regression line and φ is the concentration (expressed as volume fraction) of methanol. Linear relationships between $\log k$ and φ values were found for all compounds in the range of eluent composition examined ($r \geq 0.99$). Possible occurrence of retention changes due to column aging was monitored by checking the retention times of test compound (ibuprofen). During the study, retention time of test compound changed no more than 3.2%, and no correction was done to the retention times experimentally determined for the analytes. All $\log k$ values are the average of three measurements, with RSD values lower than 4.4%; the 95% confidence interval associated with each $\log k_{w\ C18}$ value never exceeded 0.25 (Table 1).

The results indicate that both experimental chromatographic methods yielded similar data. Positive hydrophobicity parameters were found for all analytes. The highest hydrophobicity was found for azathioprine ($R_{Mw} = 1.41$; $\log k_{w\ C18} = 1.34$), while 6-thioguanine had the lowest affinity for C18 stationary phase ($R_{Mw} = 0.21$; $\log k_{w\ C18} = 0.32$). The corresponding linear correlation between chromatographic hydrophobicity parameters featured $r = 0.93$. Although limited data on hydrophobicity of immunosuppressants are available in literature, it should be pointed out that comparable data were obtained by various techniques [42–44].

2.1.2. Phospholipid Binding Assay

As already reported in literature, the octadecylsilane stationary phase does not resemble a biomembrane phospholipid bilayer and consequently does not fully capture the complex and dynamic nature of human intestinal absorption of drugs [45]. Immobilized artificial membrane (IAM) columns, on the other hand, have shown satisfying performance regarding this matter [46]. In an attempt to gain insight into interaction between azathioprine, 6-mercaptopurine, 6-thioguanine, folic acid and cell membrane, IAM chromatographic column with a monolayer of phosphatidylcholine covalently bound to a propylamino-silica core was used. IAM column permitted the use of phosphate-buffered saline without addition of organic modifier, leading to directly measured $\log k_{w\ IAM}$ values and reducing

Simultaneous analytical methods for determination of active ingredients in FDC should be developed in advance, since these methods are required for finished product quality control monitoring and dissolution testing. The review of literature reveals that there are many analytical methods for determination of immunosuppressants and folic acid individually in pharmaceutical dosage forms. In the last decade, various techniques for determination of azathioprine have been used, including high performance thin-layer chromatography [13], spectrophotometry [14], atomic absorption spectrometry [15], flow injection chemiluminometry [16], electrochemistry [17–19] and high-performance liquid chromatography [20]. Several methods have been reported for determination of 6-mercaptopurine in pharmaceutical formulations using analytical techniques such as Raman spectroscopy [21], electrochemistry [22] and high-performance liquid chromatography [23]. In literature, several analytical methods to quantify 6-thioguanine using fluorescence spectroscopy [24], electrochemistry [25] and high performance liquid chromatography with post column iodine-azide reaction [26] can be found. Many methods for determination of folic acid have been developed due to its biological significance. These methods include fluorescence spectroscopy [27]; spectrophotometry [28]; electrochemistry [29–31]; high-performance liquid chromatography coupled with various detectors, such as diode-array detector (DAD) [32]; tandem mass spectrometry [33] and corona-charged aerosol [34] detector and ultra-performance liquid chromatography [35]. Determination of these compounds in combination with other drugs by various techniques has been reported. The most commonly used technique is electrochemistry [36–39], while less prevalent are high-performance liquid chromatography [40] and electrophoresis [41]. To the best of our knowledge, no method for simultaneous determination of all these compounds was proposed.

In light of the foregoing considerations, in this work pharmacokinetic properties of thiopurine immunosuppressants and folic acid were systematically evaluated using biomimetic chromatography. In addition, an analytical method for their simultaneous determination in proposed dose ratio as a prerequisite for development of three FDCs containing folic acid and one of thiopurine immunosuppressants is proposed.

2. Results and Discussion

2.1. Pharmacokinetic Profiling

2.1.1. RP-TLC and RP-HPLC Assays

In reversed-phase chromatography, retention of analytes is governed by hydrophobicity, and this parameter was evaluated using both TLC and HPLC techniques. Due to high affinity of investigated compounds for octadecylsilane chemically bonded to porous silica applied on chromatographic plates, several analyses with mixtures of phosphate-buffered saline with various concentrations of methanol as mobile phase were performed. Methanol was chosen as the most suitable organic modifier, since it does not disturb the hydrogen bonding network of water. Although TLC technique has limited automation capabilities compared to other chromatographic techniques, its major advantage is the possibility for simultaneous determination of hydrophobicity of several compounds on the same chromatographic plate. R_M values were obtained according to Equation (1):

$$R_M = \log\left(\frac{1}{R_F} - 1\right), \tag{1}$$

where R_F represents the retention factor. Furthermore, chromatographic parameter of hydrophobicity, R_{Mw}, was obtained by extrapolation to pure buffer using the linear Equation (2):

$$R_M = R_{Mw} - S\Phi, \tag{2}$$

where S is the slope of the regression line and φ is the concentration (expressed as volume fraction) of methanol. Linear relationships between R_M and φ values were found for all compounds in the range acid, according to European Crohn and Colitis Organization (ECCO) guidelines, should be recommended in all IBD patients in anticipation of pregnancy. Several studies have shown that some patients need to be on a higher dose of folic acid than the ECCO recommended dose [6]. Medication compliance has shown to be an important challenge in the treatment of IBD patients as the effective therapy requires long-term management. The compliance rate in IBD patients is not high (up to 60% of patients) [7]. The compliance rate in IBD patients is even higher (up to 60% of patients in remission). During this period, patients often forget or deliberately avoid taking medications at the scheduled time [8]. For diseases that require treatment with multiple drugs, such as CD and UC, fixed-dose combinations (FDC) therapy in which two or more active pharmaceutical ingredients (APIs) are combined in a single dosage form, can help in addressing some of the problems of compliance. Strategies for the development of FDCs are commonly primarily based on drug–drug interactions, the therapeutic effect of drug combinations, the prolonged period of drug action [9,10]. Although the number of marketed FDC products has grown in the last years and although the number of marketed FDC products is quite a few and the risk of too high complexities in these products is a priority of FDC. The reason for such a high complexity is the fact that the majority of FDC products contain properties. To cope with this issue, the rational strategy for development of profiles of APIs and evaluation of pharmacokinetic profiles of APIs as well as development of a delivery method for determination of APIs in early phases of FDC development [9,11].

A vast majority of IBD patients are effectively treated with thiopurine immunosuppressants (azathioprine, 6-mercaptopurine and 6-thioguanine) as steroid-sparing therapy in combination with folic acid to improve the outcomes of medical treatment (Figure 1). Still, due to limited research and development of FDC in IBD, these products are not yet available at the market.

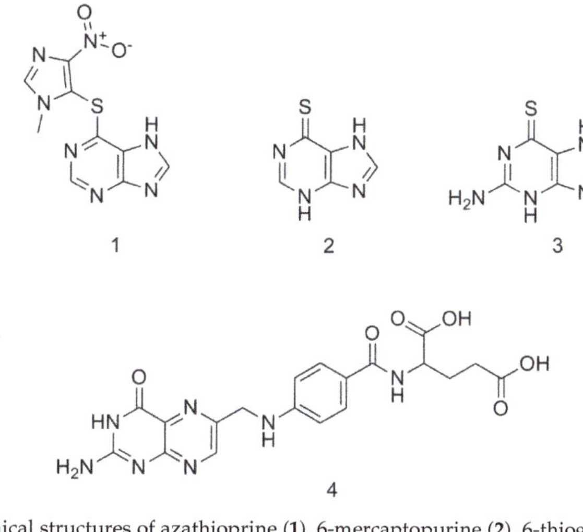

Figure 1. Chemical structures of azathioprine (**1**), 6-mercaptopurine (**2**), 6-thioguanine (**3**) and folic acid (**4**).

Biomimetic chromatography has unfolded new perspectives for the use of chromatographic techniques in drug development as it provides simple, reliable and inexpensive measurements of drug affinity to phospholipids and human proteins. The availability of this data in early stages of FDC development is crucial for a rational selection of drug combination, their pharmacokinetic profiling and finally formulation development. To our knowledge, limited data on biomimetic

Article

Pharmacokinetic Profiling and Simultaneous Determination of Thiopurine Immunosuppressants and Folic Acid by Chromatographic Methods

Edvin Brusač [1], Mario-Livio Jeličić [1], Daniela Amidžić Klarić [1], Biljana Nigović [1], Nikša Turk [2], Ilija Klarić [3] and Ana Mornar [1,*]

1. Faculty of Pharmacy and Biochemistry, University of Zagreb, A. Kovačića 1, 10000 Zagreb, Croatia; ebrusac@pharma.hr (E.B.); mljelicic@pharma.hr (M.-L.J.); damidzic@pharma.hr (D.A.K.); bnigovic@pharma.hr (B.N.)
2. Clinical Hospital Center Zagreb, Kišpatićeva 12, 10000 Zagreb, Croatia; niksa_turk@net.hr
3. Public Health Brčko DC, R. Dž. Čauševića 1, 76100 Brčko DC, Bosnia and Herzegovina; klaric67@gmail.com
* Correspondence: amornar@pharma.hr; Tel.: +385-1-481-8288

Received: 29 August 2019; Accepted: 22 September 2019; Published: 24 September 2019

Abstract: With the increase in the number of medicines patients have to take, there has been a rapid rise of fixed-dose combinations (FDCs) in the last two decades. Prior to FDC development, pharmacokinetic properties of active pharmaceutical ingredients (APIs) have to be evaluated, as well as methods for their determination developed. So as to increase patient compliance in inflammatory bowel disease, three novel FDCs of thiopurine immunosuppressants and folic acid are proposed; physico-chemical and pharmacokinetic properties such as hydrophobicity, lipophilicity and plasma protein binding of all APIs are evaluated. Moreover, experimental results of different properties are compared to those computed by various on-line prediction platforms so as to evaluate the viability of the in silico approach. A simultaneous method for their determination is developed, optimized, validated and applied to commercial tablet formulations. The method has shown to be fast, selective, accurate and precise, showing potential for reliable determination of API content in proposed FDCs during its development.

Keywords: inflammatory bowel disease; fixed-dose combination; biomimetic chromatography; thiopurine immunosuppressants; folic acid

1. Introduction

Inflammatory bowel disease (IBD) comprises a group of idiopathic inflammatory gastrointestinal diseases with intestinal and extraintestinal manifestations. Crohn's disease (CD) and ulcerative colitis (UC) are two main phenotypes of IBD; conditions that are chronic and often progressive with periods of exacerbation and remission. Despite the wide range of biological agents arising in IBD each year, the cornerstone of IBD treatment are still classical immunosuppressants (methotrexate, thiopurines), corticosteroids in flares and 5-aminosalicylates predominantly in UC [1]. Interventions with antibiotics, such as ciprofloxacin and corticosteroids, are often required in addition to maintenance therapy [2]. Moreover, IBD patients are at a risk for a variety of nutritional deficiencies because of reduced nutrient intake or absorption, as well as increased nutrient losses. Numerous nutrient deficiencies have been reported in IBD patients with varying degrees of prevalence and clinical significance. Many studies have shown that folate deficiency is one of the most noteworthy in CD and UC patients. It is common in up to 80% of IBD patients and is associated with megaloblastic anemia and increased risk of developing colorectal cancer [3–5]. Moreover, folic acid, according to European Crohn and Colitis Organization (ECCO) guidelines, should be recommended in all IBD patients in anticipation of pregnancy. Several the time of analysis considerably. All values of log $k_{w\ IAM}$ are the average of three measurements, with RSD values lower than 4.1%. Possible occurrence of retention changes due to column aging was monitored by checking the retention times of test compound (mesalazine). During the study the retention time of test compound did not change markedly (RSD = 4.3%), and no correction was done to the retention time experimentally determined for the analytes.

Generally, investigated compounds have shown lower affinity for IAM than for C18 stationary phase (Table 1). The obtained values were in the range from −0.26 (folic acid) to 0.68 (azathioprine). The lack of significant correlation between log $k_{w\ IAM}$ and log $k_{w\ C18}$ values (r = 0.61) as well as log $k_{w\ IAM}$ and R_{Mw} values (r = 0.85) confirms that partition in phospholipids encodes not only hydrophobic intermolecular recognition forces but also ionic bonds, due to electrostatic interactions between electrically charged species and phospholipids. The highest difference between log $k_{w\ C18}$ and log $k_{w\ IAM}$ values (1.25) was found for folic acid, the only compound with two carboxylic groups on the L-glutamate part of the molecule. A possible explanation for the higher affinity of neutral forms of azathioprine, 6-mercaptopurine and 6-thioguanine for IAM surface compared to negatively charged folic acid at pH 7.4 is that repulsive forces between two carboxylic groups and IAM phosphates have occurred.

2.1.3. Protein Binding Assays

As co-administered drugs may exert a competition for the same binding site on plasma proteins, and this competition can have potential clinical consequences, it is of essential importance to evaluate interaction of drugs with plasma proteins in early phases of FDC development. Chromatographic columns with HSA- and AGP-coated stationary phases have been initially developed for chiral separation. Several studies have shown that the retention times of compounds on these phases are proportional to the dynamic distribution constants of the compound between the mobile and stationary phases. It follows that these two biomimetic columns may be used in drug-plasma protein binding studies [45,47,48]. To obtain data suitable for comparison with binding data obtained by other commonly used techniques as well as for inter-laboratory comparison, a standardization and validation of biomimetic chromatographic methods should be done using a set of compounds for which the percentage of binding data is available in literature. In this study, the gradient retention times are standardized using a calibration set of compounds with known percentage of plasma protein binding data (PPB_{lit}). The known PPB_{lit} values were converted to the affinity constant, log K_{PPB} using Equation (5):

$$\log K_{PPB} = \log \frac{PPB_{lit}}{101 - PPB_{lit}}. \quad (5)$$

Acceptable correlations were found between the literature plasma protein binding data and the experimental binding data expressed as gradient retention times (t_G) for the validation set of compounds, with Equations (6) and (7) describing the dependence of log K values on t_G for HSA and AGP columns, respectively:

$$\log K_{HSA} = 1.455\ (\pm 0.246) \log t_{G\ HSA} + 0.172\ (\pm 0.120) \\ n = 10,\ r = 0.98,\ s_E = 0.159 \quad (6)$$

$$\log K_{AGP} = 5.004\ (\pm 1.384) \log t_{G\ AGP} - 0.964\ (\pm 0.350) \\ n = 10,\ r = 0.95,\ s_E = 0.278, \quad (7)$$

where the values in brackets represent 95% confidence intervals, while s_E represents the standard error. Furthermore, using the slope and intercept values from the calibration line, the logarithmic retention times were converted to log K_{PPB} values that can be converted to PPB_{HSA} and PPB_{AGP} of investigated compounds using Equation (8):

$$PPB_{exp} = \frac{101 \times 10^{\log K_{PPB}}}{1 + 10^{\log K_{PPB}}}. \quad (8)$$

Validation of the methods was performed by evaluation of retention of warfarin enantiomers. As retention time on HSA and AGP columns is dependent on the injected amount of compound, the smallest possible amount of sample was injected so that the compound would not saturate the specific binding site. Wide peaks with tailing were observed on both protein stationary phases using concentrations of test compound higher than 10 mg/L. Possible occurrence of retention changes due to column aging was monitored by checking the retention times of selected test compound. During the study no retention value changed more than 0.63% (HSA column) and 0.28% (AGP column).

HSA is the major plasma protein with vital functions acting as depot and carrier for many drugs, and it is also found to be considerably distributed in the interstitial fluid of body tissues. It is well known that HSA binds neutral and negatively charged compounds more strongly than positively charged ones. Thus, the highest binding affinity was found for folic acid (69.40%), the only investigated compound that is, due to fully ionized glutamate carboxyl groups, present in a form of negatively charged folate at the physiological pH (Table 1). Relatively high affinity for HSA was found for lipophilic azathioprine (log $k_{w\ IAM}$ = 0.68, PPB_{HSA} = 49.22%), while 6-mercaptopurine (log $k_{w\ IAM}$ = −0.19, PPB_{HSA} = 20.08%) and 6-thioguanine (log $k_{w\ IAM}$ = 0.03, PPB_{HSA} = 24.11%) bound to HSA to a lesser extent. The obtained results suggest that lipophilic forces are generally predominant in the immunosuppressive drugs–HSA binding mechanism.

Compared to HSA, AGP has a lower plasma concentration in humans. Still, the concentration of AGP in plasma can significantly increase in various conditions, such as inflammatory diseases. All immunosuppressants have shown greater extent of binding to AGP than folic acid (at least 18.7% higher) (Table 1). Given that basic drugs have been shown to bind preferentially to AGP, a better understanding of interaction between this protein and immunosuppressants is valuable. As in the case of HSA-binding mechanism, lipophilic azathioprine showed the highest percentage of AGP binding (log $k_{w\ IAM}$ = 0.68, PPB_{AGP} = 75.96%), followed by 6-thioguanine (log $k_{w\ IAM}$ = 0.03, PPB_{AGP} = 45.96%) and 6-mercaptopurine (log $k_{w\ IAM}$ = −0.19, PPB_{AGP} = 22.15%). The noticeable difference in AGP binding between structurally similar 6-thioguanine and 6-mercaptopurine is likely due to the presence of the basic amino group at position C2 of 6-thioguanine.

2.1.4. Comparison of Biomimetic Chromatographic and Computational Data

In silico based pharmacokinetic modeling approaches have been widely used in drug development processes to provide a fast, low-cost and preliminary screening of compounds prior to in vitro testing [45,49,50]. To perform a comprehensive evaluation of in silico potential in drug development, a number of compounds should be investigated. Still, due to limited number of immunosuppressants used in treatment of IBD available at market, as well as therapeutically reasonable candidates for FDC, a limited number of compounds was included in this study. The collected biomimetic chromatographic data were compared to selected physico-chemical and pharmacokinetic properties obtained by 15 established computational medicinal chemistry methods. Observed parameters included partition coefficient, human intestinal absorption, human colorectal carcinoma cells (Caco-2) permeability and plasma protein binding. The list of theoretical approaches and calculated parameters is given in Table 2.

In every respect, noticeable discrepancies among theoretical partition coefficients were observed. Intercorrelation between the data gave correlation coefficients in the range from 0.01 to 0.99. Average calculated log P values were in the range from −0.84 to 1.22. It is obvious that some of calculation procedures overestimate log P values of investigated compounds, while others underestimate them. Still, it can be found that MlogP values correlate well with both hydrophobic chromatographic parameters featuring r = 0.93 with R_{Mw} and r = 0.99 with log $k_{w\ C18}$. As mentioned above, the interaction of investigated compounds with membrane phospholipids is complex and governed not only by hydrophobicity but also electrostatic interactions, especially in the case of negatively charged folic acid. Therefore, it is worth pointing out that the linear correlation between log $k_{w\ IAM}$ and log P values calculated by programs Mcule and Chemicalize is noteworthy featuring $r \geq 0.94$, while correlations with other predicted log P values were lower ($0.01 \leq r \leq 0.85$).

Table 2. List of calculated physico-chemical and pharmacokinetic parameters.

Parameter	Program	6-thioguanine	6-mercaptopurine	Folic Acid	Azathioprine
partition coefficient	ACDLabs	−0.12	−0.18	−2.48	0.67
	ADMETLab	0.60	1.02	−0.05	1.15
	ALOGPS	−0.36	−0.13	−0.04	0.84
	ChemExper	−0.02	0.43	−1.43	0.42
	Chemicalize	−0.35	−0.12	−0.69	1.17
	Mcule	1.18	1.02	1.00	1.67
	MlogP	−1.31	−1.16	−0.62	−0.26
	Molinspiration	−0.58	−0.39	−2.37	0.50
	Molsoft	−0.36	0.23	−0.95	0.10
	pkCSM	0.60	1.02	−0.04	1.15
	PreADMET	−0.35	−0.47	0.65	0.67
	Silicos-it	2.05	2.78	0.24	−1.11
	SwissADME	0.28	0.62	−0.42	0.04
	XlogP	−0.07	0.70	−1.08	0.10
human intestinal absorption (%)	ADMETLab	76.7	81.3	39.8	78.2
	pkCSM	91.0	91.0	2.3	78.6
	PreADMET	83.1	87.8	23.2	75.5
Caco-2 permeability [log(cm × 10^{-6}/s)]	ADMETLab	0.82	1.53	−0.33	1.41
	admetSAR	0.49	0.62	−1.07	0.11
	pkCSM	0.17	1.14	−0.98	0.62
	PreADMET	0.29	0.28	0.31	−1.38
plasma protein binding (%)	ADMETLab	27.0	24.2	67.9	35.8
	pkCSM	32.2	28.1	63.9	60.4

The obtained results might indicate that theoretical approaches of the two above-mentioned programs are powerful enough to predict affinity of investigated compounds for immobilized phosphatidylcholine.

Finally, it is relevant to evaluate whether IAM, HSA and AGP biomimetic chromatographic columns might be comparable with in silico pharmacokinetic profiling of selected drugs. As uptake of folic acid is mediated by transport proteins, it was excluded from evaluation of passive absorption by IAM chromatography. Evaluation of all indices describing drug absorption indicates that the majority of the programs predict somewhat higher absorption rates for 6-mercaptopurine and 6-thioguanine compared to more lipophilic azathioprine. This might be related to their smaller size and thus easier penetration through the cell membrane. It might be concluded that passive absorption of the drug through the gastrointestinal barrier is a very heterogeneous process that cannot be accounted for via simple affinity for immobilized phosphatidylcholine, but other properties, such as the size of the molecule, should also be considered.

The trend of increasing affinity for immobilized human serum albumin column in order 6-mercaptopurine < 6-thioguanine < azathioprine < folic acid was in accordance with both parameters describing plasma protein binding calculated by ADMETLab and pkCSM programs. Moreover, it should be emphasized that the linear correlation between PPB_{HSA} values and computationally predicted values featured r values higher than 0.94. On the other hand, experimental AGP binding data did not relate well to any of predicted parameters ($r \leq 0.52$).

It should be pointed out that none of calculation procedures clearly state which plasma protein is involved in distribution of investigated compounds, which leaves room for considerable improvement of theoretical approaches.

2.2. Analysis of Pharmaceutical Dosage Forms

2.2.1. Method Development

Due to different physico-chemical properties of immunosuppressants and folic acid, as well as high dose differential between APIs combined in three proposed FDCs (5 mg of folic acid combined with 40 mg of 6-thioguanine or 50 mg of azathioprine or 50 mg of 6-mercaptopurine per dosage form), developing a unique sample preparation procedure and simultaneous chromatographic method for their determination in FDC was quite a challenging task. Optimization of the experimental conditions was split into two major areas. First, simultaneous chromatographic method for determination of active ingredients in FDC was optimized. Afterwards the proposed method was used for optimization of sample preparation procedure.

In the light of available individual methods for the determination of azathioprine, 6-mercaptopurine, 6-thioguanine and folic acid, different chromatographic parameters were tested to establish the most suitable chromatographic conditions for their simultaneous determination in FDCs [20,23,32]. Parameters such as composition of stationary and mobile phases as well as temperature of analysis were systematically studied so as to achieve good resolution of adjacent peaks, symmetric peaks, high column performance and acceptable runtime. Afterwards, extraction solvent and time were optimized to attain high extraction efficiency.

A type of chosen chromatographic column had a significant influence on separation of compounds. Due to considerable differences in polarity of analytes, a compromise between hydrophilic and lipophilic character of sorbents had to be found [20,23,32,33]. Therefore, three kinds of chromatographic columns were investigated: XBridge C18 (150 × 4.6 mm, particle size 5 μm) by Waters (Milford, MA, USA), XBridge Phenyl (150 × 4.6 mm, particle size 5 μm) by Waters (Milford, MA, USA) and Zorbax SB C8 (150 × 4.6 mm, particle size 5 μm) by Agilent Technologies (Santa Clara, CA, USA). The baseline separation of critical pair of analytes, 6-thioguanine and 6-mercaptopurine, was achieved only using Zorbax SB C8 column, which was shown to be the most suitable for the separation of analytes in a single run and this column was used for all further experiments.

The composition of mobile phase appeared to be another critical factor in achieving the appropriate chromatographic behavior and satisfying separation of analytes. The selection of acetonitrile as organic modifier, as opposed to methanol, was based primarily on lower viscosity and, accordingly, lower operating pressure. It was found that pH of mobile phase had a profound effect on the ionization of analytes and the resultant chromatographic behavior, especially in the case of folic acid. The pK_a's of two carboxylic groups on the L-glutamate part of folic acid are 3.5 and 4.8, respectively [33,40]. Therefore, under reversed phase conditions an acidified mobile phase was required to improve retention of folic acid. To optimize retention as well as peak shape of folic acid, discrete levels (0.05% to 0.2%) of a volatile organic acid (acetic and formic acid) were added to the mobile phase. Peak tailing was noted with both investigated organic acids' concentrations of less than 0.1% (peak asymmetry factor was greater than 1.2). Addition of both organic acids to mobile phase at levels 0.1% and 0.2% was found to give symmetric peaks, higher column performance and good resolution among all four analytes. Mobile phase with 0.1% of formic acid was selected for further investigations by considering the working life of the column. Initially, isocratic mode of separation was tested and was found insufficient to resolve all four analytes with good peak shape characteristics. Complete separation was achieved by conversion of the isocratic elution to a gradient mode. Moreover, the optimized gradient program dramatically facilitated achievement of shorter runtime (all analytes were eluted within 7 min).

As the column temperature may affect the efficiency of chromatographic separation, the impact of three different temperatures (20, 25 and 30 °C) using the retention and resolution factors as the basic criteria was evaluated. Inadequate peak symmetry (less than 0.7) was observed at 20 °C. The separation was markedly faster at higher temperatures, but seeing that 6-thioguanine eluted considerably earlier, resolution between 6-thioguanine and the adjacent peaks of excipients became critical. Thus, the temperature of 25 °C was selected as the optimal temperature. Different compounds

(6-methylthioguanine, methotrexate and caffeine) were assessed to be used as an internal standard. Among these compounds, 6-methylthioguanine showed good resolution and instrumental response with all APIs included in three proposed FDCs and was therefore selected as the internal standard. A chromatogram of mix standard solution separated by optimized chromatographic method is shown in Figure 2A.

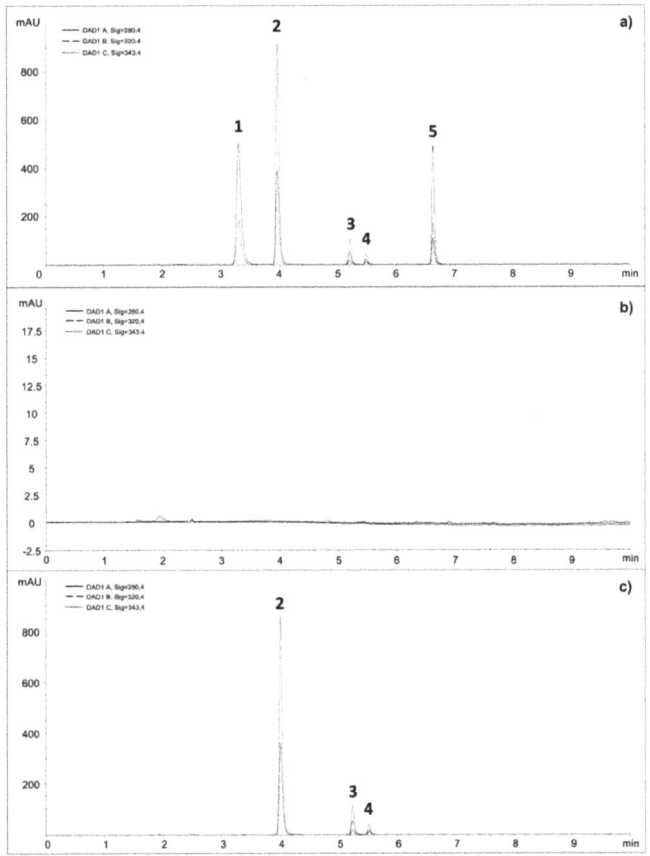

Figure 2. Chromatograms of (a) mixed standard solution of 6-thioguanine (1), 6-mercaptopurine (2), internal standard (3), folic acid (4) and azathioprine (5); (b) 0.02 M NaOH solution containing commonly used excipients; (c) proposed fixed-dose combination (FDC) containing 6-mercaptopurine (50 mg/L) (2) and folic acid (5 mg/L) (4) with the addition of internal standard (3).

Differences in chemical properties of the studied compounds hindered the simultaneous extraction of all four compounds using one set of extraction conditions. Protocols utilizing aqueous or organic solvent extraction solutions (acidic, alkaline or neutral pH), combined with at ambient and elevated temperatures shaking and sonication procedures have all been reported [51]. Therefore, one-variable-at-a-time method was applied to find a ubiquitous method suitable for all investigated analytes. Preliminary investigations have shown that methanol improves extraction of 6-thioguanine, 6-mercaptopurine and azathioprine (it was found to be more than 5.7% effective in methanol than in ultrapure water), while folic acid was found slightly soluble in organic solvents (extraction efficiency in methanol was only 3.49%) as well as in water (extraction efficiency was 22.4%), but soluble in basic aqueous solutions. Therefore, a basic solution with NaOH was selected for further investigation.

Table 3. Best studied compounds.

Analyte	Parameter (Mean Value ± Standard Deviation), $n = 7$						
	t_R [1]	Rt_R [2]	R_S [3]	k [4]	N [5]	TF [6]	A [7]
6-thioguanine	3.22 ± 0.01	0.76 ± 0.001	/	1.48 ± 0.01	8716 ± 198	1.35 ± 0.02	3522.4 ± 35.5
6-mercaptopurine	3.96 ± 0.01	/	5.20 ± 0.16	2.25 ± 0.01	13371 ± 1169	1.34 ± 0.02	5333.6 ± 1169
Folic acid	5.20 ± 0.01	1.05 ± 0.001	3.37 ± 0.03	2.28 ± 0.01	79297 ± 1956	1.34 ± 0.02	388.2 ± 2.8
Azathioprine	6.63 ± 0.02	1.275 ± 0.001	14.22 ± 0.07	3.14 ± 0.01	110071 ± 2807	1.34 ± 0.02	141.8 ± 1.7
							1578.5 ± 16.8

[1] t_R—retention time; [2] Rt_R—relative retention time; [3] R_S—resolution; [4] k—capacity factor; [5] N—theoretical plates; [6] TF—USP tailing factor; [7] A—peak area; /—not applicable.

The selectivity of the developed method was evaluated by injecting a mixture of standard

Evaluation of optimal concentrations of NaOH (0.02 M, 0.05 M and 0.1 M) showed that extraction of all analytes has been improved in 0.02 M NaOH, while degradation of azathioprine was observed at higher levels of NaOH. Further, the investigation of ultrasonication time (up to 120 min) indicated that higher extraction rates (more than 96.1%) and better precisions (RSD lower than 3.3%) for all compounds were achieved when the time was 15 min or longer. Folic acid in solution is known to decompose when exposed to ultraviolet light and/or increased temperature; thus, all experiments were performed at ambient temperature using amber laboratory glassware [52].

2.2.2. Validation of the Method

The newly developed method was validated according to the ICH guidelines with respect to selectivity, linearity, sensitivity, precision, accuracy, stability and robustness [53]. All validation parameters were evaluated using placebo solution spiked with standard solutions.

As an integral part of the analytical method, system suitability test was used to verify adequacy of the resolution and reproducibility of the chromatographic system. It was established by analysis of seven replicate injections of mixed standard (100% level of concentration in proposed FDCs) solution and parameters such as retention time, relative retention time, resolution, capacity factor, number of theoretical plates, tailing factor and peak area of the analytes were calculated. The results of system suitability are shown in Table 3.

Table 3. System suitability data.

Analyte	Parameter (Mean Value ± Standard Deviation), $n = 7$						
	t_R [1]	Rt_R [2]	Rs [3]	K [4]	N [5]	TF [6]	A [7]
6-thioguanine	3.29 ± 0.01	0.635 ± 0.001	/	1.06 ± 0.01	8716 ± 198	1.38 ± 0.02	2871.6 ± 27.8
6-mercaptopurine	3.96 ± 0.01	0.761 ± 0.001	5.69 ± 0.05	1.48 ± 0.01	28462 ± 330	1.35 ± 0.02	3522.4 ± 35.5
6-methylthioguanine	5.20 ± 0.01	/	13.57 ± 0.31	2.25 ± 0.01	53336 ± 1169	1.34 ± 0.01	388.2 ± 2.8
Folic acid	5.49 ± 0.01	1.055 ± 0.001	3.37 ± 0.03	2.43 ± 0.01	75297 ± 1956	1.28 ± 0.01	141.8 ± 1.7
Azathioprine	6.63 ± 0.02	1.275 ± 0.001	14.22 ± 0.07	3.14 ± 0.01	110071 ± 2807	1.34 ± 0.02	1578.5 ± 16.8

[1] t_R—retention time; [2] Rt_R—relative retention time; [3] Rs—resolution; [4] k—capacity factor; [5] N—theoretical plates; [6] TF—USP tailing factor; [7] A—peak area; /—not applicable.

The selectivity of the developed method was evaluated by injecting a mixture of standard solutions and was assessed by peak purity test (comparison between analyte peak and auto threshold in the purity plot). The peaks of analyte of interest were not found to be attributed to more than one component, indicating the method to be selective (peak purity factors were higher than 999.8). Afterwards, a 0.02 M solution containing excipients chosen as the ones used commonly in tablet formulations was analyzed. The majority of the excipients were eluted from the column within the first 2.5 min, which indicates that the excipients were not interfering with the assay (Figure 2B). Finally, mix standard solution containing excipients chosen as the ones used commonly in tablet formulations was analyzed. The obtained peak purity factors were also higher than 999.8.

The linearity of the method was evaluated by analyzing at least six freshly prepared solutions, in the concentration range between 0.2 and 20 mg/L for folic acid, 5 and 80 mg/L for 6-thioguanine and 5 and 100 mg/L for azathioprine and 6-mercaptopurine (Table 4). Calibration curves showed linear responses for all analytes over dynamic ranges, and the corresponding regression correlation coefficients (r) were all above 0.999. Due to a high dose differential between APIs combined in proposed FDCs, limits of detection (LOD) and quantitation (LOQ) of the method were one of the most important validation parameters. LOD and LOQ, based on 3 and 10 times the signal-to-noise ratio, respectively, were determined by repeated injections of diluted standard solutions. According to data presented in Table 4, satisfying linearity range as well as sensitivity was obtained for all four analytes.

Table 4. Method calibration data.

Analyte	Linearity Range (mg/L)	Equation	r [1]	s_E [2]	LOD (mg/L)	LOQ (mg/L)
6-thioguanine	5–80	$y = 0.251\,x - 0.132$	0.9996	0.124	0.1	0.2
6-mercaptopurine	5–100	$y = 0.230\,x - 0.034$	0.9997	0.140	0.1	0.2
Folic acid	0.2–20	$y = 0.081\,x - 0.007$	0.9999	0.013	0.1	0.2
Azathioprine	5–100	$y = 0.102\,x + 0.016$	0.9998	0.048	0.1	0.2

[1] r—Pearson correlation coefficient; [2] s_E—standard error.

Precision studies were done by evaluation of repeatability and intermediate precision. Intra-day studies were done by injecting the solutions at one concentration level (100% level of concentration in proposed FDCs) six different times on the same day. Inter-day studies were done in triplicate every day up to three consecutive days. Low values of RSD (%) showed that the method is precise within the acceptance limit of ±2%. The intra- and inter-day variability or precision data are given in Table 5. The results indicate good precision of the developed method.

Table 5. Intra- and inter-day assay precision and accuracy data.

Analyte	Precision Relative Standard Deviation (RSD) (%)		Accuracy Recovery ± RSD (%)		
	Intra-Day Precision ($n = 6$)	Inter-Day Precision ($n = 9$)	Low	Medium	High
6-thioguanine	0.73	0.75	99.49 ± 0.80	100.42 ± 1.66	98.79 ± 0.68
6-mercaptopurine	0.81	0.79	100.35 ± 1.74	100.79 ± 0.69	99.01 ± 0.76
Folic acid	1.04	1.11	99.78 ± 0.81	97.57 ± 0.37	100.25 ± 0.87
Azathioprine	0.40	0.64	100.38 ± 0.86	100.38 ± 1.19	99.55 ± 0.58

The proposed method was evaluated for its accuracy by the analysis of the placebo solutions spiked with three different standard concentration levels (80%, 100% and 120% level of concentration in proposed FDCs). As it can be seen in Table 5, the calculated accuracy was always within the acceptance limit of ±3% of the nominal concentration.

The stability of APIs' sample solutions was evaluated by keeping them in tightly closed amber vials on the laboratory worktable at room temperature for 2 h and in a rack on the autosampler at 4 °C for 24 h. Furthermore, long-term stability was evaluated in the freezer at −20 °C for 30 days. The recoveries were in the range from 96.4% to 103.3%, which indicates that investigated compounds were stable in described experimental conditions within the given periods.

The impact of different chromatographic parameters on the peak areas and peak shapes of all APIs was examined by small deliberate changes in the mobile phase composition (±1%) and flow rate (±10%) as well as temperature (± 1 °C) to establish the robustness of the proposed method. In all deliberately varied conditions, the RSD of peak areas of all APIs were found to be well within the acceptable limit of 5%. The tailing factor for all peaks was found to be lower than 1.5.

2.2.3. Application of the Method

The developed method was successfully applied for analysis of azathioprine, 6-mercaptopurine, 6-thioguanine and folic acid in marketed tablet formulation. Firstly, 20 tablets of each pharmaceutical product were weighed individually; the RSDs of the tablet weights were lower than 3.0%, indicating satisfying weight uniformity. Upon analysis of the samples, no interferences of excipients were observed as peak purity factors were higher than 999.7. The amounts recovered were expressed as a percentage of the label claim, and obtained results were in the range between 96.1% and 104.4% (Table 6). Contents of all of the analyzed samples conform to the United States Pharmacopoeia [54] and British Pharmacopoeia regulations [55].

Table 6. Results of analyses of marketed formulations ($n = 3$).

Commercial Formulation	Manufacturer	Active Substance	Amount Labeled (mg)	Amount Found (mg)	Detected/ Labeled (%)	RSD (%)
Thiosix®	Teva Nederland B.V, Haarlem, the Netherlands	6-thioguanine	10	10.44	104.42	2.90
Puri-Nethol®	Aspen, Dublin, Ireland	6-mercaptopurine	50	48.07	96.13	2.89
Folacin®	JGL, Rijeka, Croatia	folic acid	5	4.99	99.95	3.31
Imuran®	Aspen, Dublin, Ireland	azathioprine	50	51.63	103.26	2.24

Afterwards, the method was successfully applied to analysis of all three proposed FDCs. A representative chromatogram for analysis of proposed FDC containing 6-mercaptopurine and folic acid is shown in Figure 2C.

3. Materials and Methods

3.1. Reagents and Chemicals

Standards, certified reference material, for calibration and validation of biomimetic columns were: Acetylsalicylic acid, caffeine, cefalexin, chloramphenicol, cimetidine, codeine hydrochloride dihydrate, furosemide, ibuprofen, mesalazine, metronidazole, nifedipine, paracetamol, propranolol hydrochloride, salicylic acid, sulfadiazine, sulfamethoxazole, thiamine hydrochloride and warfarin purchased from Sigma-Aldrich (St. Louis, MO, USA). Azathioprine; folic acid; 6-mercaptopurine and 6-thioguanine, United States Pharmacopeia reference standard and 6-methylthioguanine ($\geq$95%) have also been obtained from Sigma-Aldrich. Acetonitrile, ethanol, iso-propanol, methanol, formic acid, all HPLC grade; sodium hydroxide; ACS reagent (97.0%) pellets; sodium nitrate ($\geq$99.0); phosphate-buffered saline tablets (0.01 M, 0.0027 M potassium chloride and 0.137 M sodium chloride, pH 7.4 at 25 °C); potassium phosphate monobasic and potassium phosphate dibasic (HPLC grade) were provided by Sigma-Aldrich. Ultra-pure water was obtained using a MilliQ UF-Plus system (Millipore, Darmstadt, Germany); resistivity MΩcm^{-1} >18 at 25 °C and TOC <5 ppb. The excipients included hydroxypropyl methylcellulose Methocel K100M Premium CR (Colorcon, Harleysville, PA, USA), magnesium stearate (Acros Organics, Princeton, NJ, USA), lactose monohydrate, stearic acid, wheat, rice and corn starch (Kemig, Zagreb, Croatia).

3.2. Equipment

All assays were performed on an Agilent 1100 series HPLC system (Agilent Technologies, Waldbronn, Germany) consisting of a quaternary pump, autosampler, vacuum degasser, column compartment and DAD. Data acquisition and processing were carried out using ChemStation for LC 3D software (version rev A.10.02[1757]).

Evaluation of TLC plates was done using Camag UV Cabinet 4 consisting of a UV Lamp 4 (dual wavelength 254/366 nm) and a Viewing Box 4 (Camag, Muttenz, Switzerland), while Elmasonic Ultrasonic BathModel XtraTT with heat and time controller (Elma Schmidbauer, Singen, Germany) was employed for all ultrasonic extractions. In sample preparation step centrifuge model Z 326 K by Hermle (Wehingen, Germany) with digital time, temperature and speed control were also used. Weighing measurements were performed using MX5 Microbalance (Mettler-Toledo, Greifensee, Switzerland) with 1 µg readability.

3.3. Preparation of Stock and Working Solutions

Solutions containing 100 mg/L of standards used for calibration and validation of biomimetic columns were dissolved in methanol. For the simultaneous determination method, stock solutions containing 200 mg/L of azathioprine, 6-mercaptopurine and 6-thioguanine were prepared by dissolving accurately weighed amounts of reference standards in diluent 1 (containing methanol and 0.1 M NaOH, 95:5, *v/v*), while stock solution containing 200 mg/L of folic acid was prepared by dissolving accurately weighed amount of reference standard in diluent 2 (containing methanol, ultrapure water and 0.1 M NaOH, 70:29:1, *v/v/v*). Working solutions of lower concentrations were freshly prepared by dilution with water. Internal standard, 6-methylthioguanine (200 mg/L in diluent 1) was added to each working solution to make its concentration 10 mg/L in each sample. For the validation study, placebo solution in 0.02 M NaOH containing mixture of the most commonly used excipients was used. The drug to excipient ratio used was similar to that in commercial formulations. All solutions were stored at 4 °C in amber glassware. Prior to use, all solutions were filtered through 0.2 μm polyethersulfone filters (Obrnuta faza, Pazin, Croatia).

3.4. RP-TLC Assay

TLC experiments were performed on commercially available 10 × 10 cm RP-18 TLC plates (Merck, Darmstadt, Germany). Methanol-phosphate-buffered saline mixtures were used as mobile phases. The methanol volume fraction was varied from 20% to 80% in 5% steps. A sufficient amount of mobile phase was placed in a Camag flat-bottom chamber for 10 × 10 cm plates with a stainless-steel lid to have a level of 5 mm. After saturation of chambers with solvent vapor for 30 min, the plates were developed to a distance of 9 cm at room temperature, dried in open air for 5 min and visualized under λ = 254 nm UV light.

3.5. RP-HPLC Assay

Hydrophobicity of the compounds was also studied on a reversed-phase Symmetry C18 column (150 × 4.6 mm, 3.5 μm particle size) obtained by Waters (Milford, MA, USA). Methanol-phosphate-buffered saline mixtures were used as mobile phases. The methanol volume fraction was varied from 10% to 50% in 5% steps. Each mobile phase was shaken vigorously, filtrated through cellulose nitrate filter (0.45 μm, Sartorius, Goettingen, Germany) and degassed by sonication (5 min) before use. Injection volume was set at 10 μL, and the measurements were carried out at 25.0 ± 0.1 °C and at flow rate 1.0 mL/min. The absorbance of the analytes during a chromatographic run was collected in the spectral range 200–400 nm, and the detection wavelength for each analyte was the one providing the maximum peak height.

3.6. Phospholipid Binding Assay

The binding of compounds to the immobilized artificial membrane (IAM) was measured using commercially available immobilized phosphatidylcholine columns (IAM.PC.DD2, 100 × 4.6 mm, 300 Å particle size) obtained by Regis Technologies (Morton Grove, IL, USA). The mobile phase consisted of phosphate-buffered saline that was filtrated through cellulose nitrate filter and degassed (5 min) by sonication before use. Injection volume was set at 10 μL and the measurements were carried out at 25.0 ± 0.1 °C and at flow rate 1.0 mL/min. The absorbance of the analytes during a chromatographic run was collected in the spectral range 200–400 nm and the detection wavelength for each analyte was the one providing the maximum peak height.

3.7. Protein Binding Assays

The interactions of analytes with plasma proteins human serum albumin (HSA) and α1-acid glycoprotein (AGP) have been measured using commercially available chemically bonded HSA (Chiralpak-HSA, 50 × 4.6 mm, 5 μm particle size) and AGP (Chiralpak-AGP, 50 × 4.6 mm, 5 μm particle size) columns obtained by ChromTech (Cedex, France). The mobile phase consisted of 20 mM potassium phosphate buffer with the pH adjusted to 7.0 (A) and iso-propanol (B). The gradient program was as follows: 0–3 min, linear gradient 0–30% B, 3–10 min, isocratic 30% B. The total run time was 15 min to allow re-equilibration of the protein phase with the buffer. Injection volume was set at 10 μL, and the measurements were carried out at 25.0 ± 0.1 °C and at flow rate 1.5 mL/min. The absorbance of the analytes during a chromatographic run was collected at the following wavelengths: 250, 280, 300, 320 and 340 nm.

3.8. Analysis of Pharmaceutical Dosage Forms

3.8.1. Sample Preparation

Twenty randomly chosen tablets were weighed, and the average weight of the tablet was determined. All tablets were crushed and powdered. For the analysis of the individual samples, an amount of powdered tablets equivalent to 0.2 mg of folic acid, 1.6 mg of 6-thioguanine, 2 mg of azathioprine or 2 mg of 6-mercaptopurine was weighed. Regarding analysis of three proposed FDCs, an adequate amount of prepared powders, equivalent to the proposed therapeutic dose ratio of each FDC (0.2 mg of folic acid combined with 1.6 mg of 6-thioguanine or 2 mg of azathioprine or 2 mg of 6-mercaptopurine) was weighed and mixed. For both of the procedures, the powders were transferred into 10 mL amber volumetric flasks and dispersed in 0.02 M NaOH with addition of internal standard solution. It was then followed by sonication for 15 min to provide complete dissolution. The final volume was adjusted with ultrapure water and the mixture was centrifuged at 6000 rpm for 10 min to separate the supernatant from non-dissolved excipients. Afterwards, the clear supernatant liquor was filtrated through a 0.2 µm polyethersulfone injection filter, appropriately diluted and transferred into an amber HPLC vial.

3.8.2. Sample Analysis

Chromatographic separation was achieved using a Zorbax SB C8 column (150 × 4.6 mm, 5 µm particle size) obtained by Agilent Technologies. A Zorbax SB C8 guard column (10 × 4.6 mm, 5 µm particle size) was also utilized. The mobile phase consisted of water (eluent A) and acetonitrile (eluent B), both acidified with 0.1% (v/v) formic acid. Each component of the mobile phase was filtrated through a cellulose nitrate filter and degassed before use in an ultrasonic bath for 5 min. The elution was a two-step gradient program with flow of 1.0 mL/min throughout the method: 0–3 min, linear gradient 5–20% B, 3–10 min linear gradient 20–70% B. Total run time was 15 min to allow re-equilibration of the stationary phase for the following analysis. Injection volume was set at 10 µL and a mixture of acetonitrile and ultrapure water (80:20, v/v) was used as needle wash solvent. The measurements were carried out at 25.0 ± 0.1 °C. The absorbance of the analytes during a chromatographic run was collected in the spectral range 200–400 nm and the detection wavelength for each analyte was the one providing the maximum peak height.

4. Conclusions

In this work, new FDCs of thiopurine immunosuppressants and folic acid are proposed. Two key prerequisites in new FDC development, pharmacokinetic profiling and simultaneous determination of APIs, have been conducted.

Evaluation of pharmacokinetic profiles of azathioprine, 6-mercaptopurine, 6-thioguanine and folic acid was performed using different chromatographic techniques and in silico approach.

The simultaneous determination method is capable of resolving and quantifying four analytes in a relatively short run time, with a simple and time-efficient sample preparation. Moreover, most commonly used excipients do not interfere with the determination of the analytes, demonstrating the selectivity of the method. The number of APIs found was in agreement with the label claim of commercially available formulations, proving the method to be reliable. The method shows potential to be used in testing of content uniformity and dissolution profiling in the formulation process.

However, proposed FDCs need to be evaluated further in terms of stability of multi-drug combination versus single API dosage forms by means of stress studies and stability-indicating methods, which will be our further goals.

Author Contributions: Conceptualization, B.N., N.T. and A.M.; Methodology, E.B. and M.-L.J.; Validation, E.B. and M.-L.J.; Formal analysis, D.A.K., I.K.; Investigation, D.A.K., I.K.; Resources, N.T. and I.K.; Writing - original draft preparation, E.B. and A.M.; Writing - Review and editing, E.B., M.-L.J., D.A.K., B.N., N.T., I.K. and A.M.; Supervision, B.N., N.T. and A.M.; Project administration, A.M.; Funding Acquisition, A.M.

Funding: This research was funded by Croatian Science Foundation, grant numbers HRZZ-UIP-2017-05-3949 and HRZZ-DOK-2018-01-9047.

Conflicts of Interest: The authors declare no conflict of interest.

References

1. Reinglas, J.; Gonczi, L.; Kurt, Z.; Bessissow, T.; Lakatos, P.L. Positioning of old and new biologicals and small molecules in the treatment of inflammatory bowel diseases. *World J. Gastroenterol.* **2018**, *24*, 3567–3582. [CrossRef] [PubMed]
2. Kabir, A.; Furton, K.G.; Tinari, N.; Grossi, L.; Innosa, D.; Macerola, D.; Tartaglia, A.; Di Donato, V.; D'Ovidio, C.; Locatelli, M. Fabric phase sorptive extraction-high performance liquid chromatography-photo diode array detection method for simultaneous monitoring of three inflammatory bowel disease treatment drugs in whole blood, plasma and urine. *J. Chromatogr. B Anal. Technol. Biomed. Life Sci.* **2018**, *1084*, 53–63. [CrossRef] [PubMed]
3. Burr, N.; Hull, M.A.; Subramanian, V. Folic Acid Supplementation May Reduce Colorectal Cancer Risk in Patients with Inflammatory Bowel Disease. *J. Clin. Gastroenterol.* **2016**, *51*, 247–253. [CrossRef] [PubMed]
4. Weisshof, R.; Chermesh, I. Micronutrient deficiencies in inflammatory bowel disease. *Curr. Opin. Clin. Nutr. Metab. Care* **2015**, *18*, 576–581. [CrossRef] [PubMed]
5. Scaldaferri, F.; Pizzoferrato, M.; Lopetuso, L.R.; Musca, T.; Ingravalle, F.; Sicignano, L.L.; Mentella, M.; Miggiano, G.; Mele, M.C.; Gaetani, E.; et al. Nutrition and IBD: Malnutrition and/or Sarcopenia? A Practical Guide. *Gastroenterol. Res. Pract.* **2017**, *2017*, 8646495. [CrossRef] [PubMed]
6. Van Der Woude, C.; Ardizzone, S.; Bengtson, M.; Fiorino, G.; Fraser, G.; Katsanos, K.; Kolacek, S.; Juillerat, P.; Mulders, A.; Pedersen, N.; et al. The Second European Evidenced-Based Consensus on Reproduction and Pregnancy in Inflammatory Bowel Disease. *J. Crohn's Coliti* **2015**, *9*, 107–124. [CrossRef] [PubMed]
7. Ghadir, M.R.; Bagheri, M.; Vahedi, H.; Ebrahimi-Daryani, N.; Malekzadeh, R.; Hormati, A.; Kolahdoozan, S.; Chaharmahali, M. Nonadherence to Medication in Inflammatory Bowel Disease: Rate and Reasons. *Middle East J. Dig. Dis.* **2016**, *8*, 116–121. [CrossRef]
8. Testa, A.; Castiglione, F.; Nardone, O.M.; Colombo, G.L. Patient Preference and Adherence Dovepress Adherence in ulcerative colitis: An overview. *Patient Prefer. Adherence* **2017**, *11*, 297–303. [CrossRef]
9. Moon, C.; Oh, E. Rationale and strategies for formulation development of oral fixed dose combination drug products. *J. Pharm. Investig.* **2016**, *46*, 615–631. [CrossRef]
10. Desai, D.; Wang, J.; Wen, H.; Li, X.; Timmins, P. Formulation design, challenges, and development considerations for fixed dose combination (FDC) of oral solid dosage forms. *Pharm. Dev. Technol.* **2013**, *18*, 1265–1276. [CrossRef]
11. Dash, R.P.; Rais, R.; Srinivas, N.R. Key Pharmacokinetic Essentials of Fixed-Dosed Combination Products: Case Studies and Perspectives. *Clin. Pharmacokinet.* **2018**, *57*, 419–426. [CrossRef] [PubMed]
12. Valko, K.L.; Teague, S.P.; Pidgeon, C. In vitro membrane binding and protein binding (IAM MB/PB technology) to estimate in vivo distribution: Applications in early drug discovery. *ADMET DMPK* **2017**, *5*, 14–38. [CrossRef]
13. Jain, P.S.; Thakre, P.; Chaudhari, A.J.; Chavhan, M.L.; Surana, S.J. Determination of azathioprine in bulk and pharmaceutical dosage form by HPTLC. *J. Pharm. Bioallied Sci.* **2012**, *4*, 318–321. [CrossRef] [PubMed]
14. Davarani, S.S.H.; Zad, Z.R.; Taheri, A.R.; Rahmatian, N. Highly selective solid phase extraction and preconcentration of Azathioprine with nano-sized imprinted polymer based on multivariate optimization and its trace determination in biological and pharmaceutical samples. *Mater. Sci. Eng. C* **2017**, *71*, 572–583. [CrossRef] [PubMed]
15. Sharma, S.; Sharma, M.C. Spectrophotometric and Atomic Absorption Spectrometric Determination and Validation of Azathioprine in API and Pharmaceutical Dosage Form. *J. Optoelectron. Biomed. Mater.* **2010**, *2*, 213–216.

16. Wang, J.; Zhao, P.; Han, S. Direct Determination of Azathioprine in Human Fluids and Pharmaceutical Formulation Using Flow Injection Chemiluminescence Analysis. *J. Chin. Chem. Soc.* **2012**, *59*, 239–244. [CrossRef]
17. Hatami, Z.; Jalali, F. Voltammetric determination of immunosuppressive agent, azathioprine, by using a graphene-chitosan modified-glassy carbon electrode. *Russ. J. Electrochem.* **2015**, *51*, 70–76. [CrossRef]
18. Vais, R.D.; Sattarahmady, N.; Karimian, K.; Heli, H. Green electrodeposition of gold hierarchical dendrites of pyramidal nanoparticles and determination of azathioprine. *Sens. Actuators B Chem.* **2015**, *215*, 113–118. [CrossRef]
19. Asadian, E.; Zad, A.I.; Shahrokhian, S. Voltammetric studies of Azathioprine on the surface of graphite electrode modified with graphene nanosheets decorated with Ag nanoparticles. *Mater. Sci. Eng. C* **2016**, *58*, 1098–1104. [CrossRef]
20. Satishchandra, M.H.; Singhvi, I.; Raj, H. Development and validation of RP-HPLC method for azathioprine in pharmaceutical dosage form. *Eur. J. Pharm. Med. Res.* **2017**, *4*, 500–503.
21. Fu, W.L.; Zhen, S.J.; Huang, C.Z. Controllable preparation of graphene oxide/metal nanoparticle hybrids as surface-enhanced Raman scattering substrates for 6-mercaptopurine detection. *RSC Adv.* **2014**, *4*, 16327–16332. [CrossRef]
22. Keyvanfard, M.; Khosravi, V.; Karimi-Maleh, H.; Alizad, K.; Rezaei, B. Voltammetric determination of 6-mercaptopurine using a multiwall carbon nanotubes paste electrode in the presence of isoprenaline as a mediator. *J. Mol. Liq.* **2013**, *177*, 182–189. [CrossRef]
23. Somasekhar, V. Optimization and validation of an RP-HPLC method for the estimation of 6-mercaptopurine in bulk and pharmaceutical formulations. *Braz. J. Pharm. Sci.* **2014**, *50*, 793–797. [CrossRef]
24. Amjadi, M.; Farzampour, L. Selective turn-on fluorescence assay of 6-thioguanine by using harmine-modified silver nanoparticles. *Luminescence* **2014**, *29*, 689–694. [CrossRef] [PubMed]
25. Smarzewska, S.; Pokora, J.; Leniart, A.; Festinger, N.; Ciesielski, W. Carbon Paste Electrodes Modified with Graphene Oxides—Comparative Electrochemical Studies of Thioguanine. *Electroanalysis* **2016**, *28*, 1562–1569. [CrossRef]
26. Zakrzewski, R. Development and validation of a reversed-phase HPLC method with post-column iodine-azide reaction for the determination of thioguanine. *J. Anal. Chem.* **2009**, *64*, 1235–1241. [CrossRef]
27. Hemmateenejad, B.; Shakerizadeh-Shirazi, F.; Samari, F. BSA-modified gold nanoclusters for sensing of folic acid. *Sens. Actuators B Chem.* **2014**, *199*, 42–46. [CrossRef]
28. Matias, R.; Ribeiro, P.R.S.; Sarraguça, M.C.; Lopes, J.A. A UV spectrophotometric method for the determination of folic acid in pharmaceutical tablets and dissolution tests. *Anal. Methods* **2014**, *6*, 3065–3071. [CrossRef]
29. Khaleghi, F.; Irai, A.E.; Sadeghi, R.; Gupta, V.K.; Wen, Y. A Fast Strategy for Determination of Vitamin B9 in Food and Pharmaceutical Samples Using an Ionic Liquid-Modified Nanostructure Voltammetric Sensor. *Sensors* **2016**, *16*, 747. [CrossRef]
30. Bandžuchová, L.; Selesovska, R.; Navratil, T.; Chylkova, J. Electrochemical behavior of folic acid on mercury meniscus modified silver solid amalgam electrode. *Electrochim. Acta* **2011**, *56*, 2411–2419. [CrossRef]
31. Cinková, K.; Švorc, Ľ.; Šatkovská, P.; Vojs, M.; Michniak, P.; Marton, M. Simple and Rapid Quantification of Folic Acid in Pharmaceutical Tablets using a Cathodically Pretreated Highly Boron-doped Polycrystalline Diamond Electrode. *Anal. Lett.* **2016**, *49*, 107–121. [CrossRef]
32. Melfi, M.T.; Nardiello, D.; Cicco, N.; Candido, V.; Centonze, D. Simultaneous determination of water- and fat-soluble vitamins, lycopene and beta-carotene in tomato samples and pharmaceutical formulations: Double injection single run by reverse-phase liquid chromatography with UV detection. *J. Food Compos. Anal.* **2018**, *70*, 9–17. [CrossRef]
33. Goldschmidt, R.J.; Wolf, W.R. Simultaneous determination of water-soluble vitamins in SRM 1849 Infant/Adult Nutritional Formula powder by liquid chromatography-isotope dilution mass spectrometry. *Anal. Bioanal. Chem.* **2010**, *397*, 471–481. [CrossRef] [PubMed]
34. Márquez-Sillero, I.; Cárdenas, S.; Valcárcel, M. Determination of water-soluble vitamins in infant milk and dietary supplement using a liquid chromatography on-line coupled to a corona-charged aerosol detector. *J. Chromatogr. A* **2013**, *1313*, 253–258. [CrossRef] [PubMed]

35. Deconinck, E.; Crevits, S.; Baten, P.; Courselle, P.; De Beer, J. A validated ultra high pressure liquid chromatographic method for qualification and quantification of folic acid in pharmaceutical preparations. *J. Pharm. Biomed. Anal.* **2011**, *54*, 995–1000. [CrossRef] [PubMed]
36. Beitollahi, H.; Raoof, J.-B.; Hosseinzadeh, R. Electroanalysis and simultaneous determination of 6-thioguanine in the presence of uric acid and folic acid using a modified carbon nanotube paste electrode. *Anal. Sci.* **2011**, *27*, 991. [CrossRef] [PubMed]
37. Martínez-Rojas, F.; Del Valle, M.A.; Isaacs, M.; Ramírez, G.; Armijo, F. Electrochemical Behaviour Study and Determination of Guanine, 6-Thioguanine, Acyclovir and Gancyclovir on Fluorine-doped SnO$_2$ Electrode. Application in Pharmaceutical Preparations. *Electroanalysis* **2017**, *29*, 2888–2895. [CrossRef]
38. Kaur, B.; Srivastava, R. Simultaneous determination of epinephrine, paracetamol, and folic acid using transition metal ion-exchanged polyaniline–zeolite organic–inorganic hybrid materials. *Sens. Actuators B Chem.* **2015**, *211*, 476–488. [CrossRef]
39. Moghaddam, H.M.; Beitollahi, H.; Tajik, S.; Soltani, H. Fabrication of a nanostructure based electrochemical sensor for voltammetric determination of epinephrine, uric acid and folic acid. *Electroanalysis* **2015**, *27*, 2620–2628. [CrossRef]
40. Jogi, K.; Rao, M.B.; Raju, R.R. Development and Validation of Stability Indicating RP-HPLC Method for the Estimation of Methotrexate and Folic Acid in Bulk and Tablet Dosage Form. *Int. J. Eng. Technol. Sci. Res.* **2016**, *3*, 45–53.
41. Maráková, K.; Piešťanský, J.; Mikuš, P. Determination of Drugs for Crohn's Disease Treatment in Pharmaceuticals by Capillary Electrophoresis Hyphenated with Tandem Mass Spectrometry. *Chromatographia* **2017**, *80*, 537–546. [CrossRef]
42. Czyrski, A.; Kupczyk, B. The Determination of Partition Coefficient of 6-Mercaptopurine Derivatives by Thin Layer Chromatography. *J. Chem.* **2013**, *2013*, 419194. [CrossRef]
43. Kowalska, A.; Pluta, K. RP TLC assay of the lipophilicity of new azathioprine analogs. *J. Liq. Chromatogr. Relat. Technol.* **2012**, *35*, 1686–1696. [CrossRef]
44. Chrzanowska, M.; Kuehn, M.; Hermann, T.; Neubert, R.H.H. Biopharmaceutical characterization of some synthetic purine drugs. *Die Pharm.* **2003**, *58*, 504–506.
45. Lasić, K.; Bokulić, A.; Milić, A.; Nigović, B.; Mornar, A. Lipophilicity and bio-mimetic properties determination of phytoestrogens using ultra-high-performance liquid chromatography. *Biomed. Chromatogr.* **2019**, *33*, e4551. [CrossRef] [PubMed]
46. Tsopelas, F.; Vallianatou, T.; Tsantili-Kakoulidou, A. The potential of immobilized artificial membrane chromatography to predict human oral absorption. *Eur. J. Pharm. Sci.* **2016**, *81*, 82–93. [CrossRef] [PubMed]
47. Chrysanthakopoulos, M.; Vallianatou, T.; Giaginis, C.; Tsantili-Kakoulidou, A. Investigation of the retention behavior of structurally diverse drugs on alpha1 acid glycoprotein column: Insight on the molecular factors involved and correlation with protein binding data. *Eur. J. Pharm. Sci.* **2014**, *60*, 24–31. [CrossRef]
48. Kamble, S.; Loadman, P.; Abraham, M.H.; Liu, X. Structural properties governing drug-plasma protein binding determined by high-performance liquid chromatography method. *J. Pharm. Biomed. Anal.* **2018**, *149*, 16–21. [CrossRef]
49. Mornar, A.; Damić, M.; Nigović, B. Pharmacokinetic Parameters of Statin Drugs Characterized by Reversed Phase High-Performance Liquid Chromatography. *Anal. Lett.* **2011**, *44*, 1009–1020. [CrossRef]
50. Russo, G.; Grumetto, L.; Barbato, F.; Vistoli, G.; Pedretti, A. Prediction and mechanism elucidation of analyte retention on phospholipid stationary phases (IAM-HPLC) by in silico calculated physico-chemical descriptors. *Eur. J. Pharm. Sci.* **2017**, *99*, 173–184. [CrossRef]
51. Blake, C.J. Analytical procedures for water-soluble vitamins in foods and dietary supplements: A review. *Anal. Bioanal. Chem.* **2007**, *389*, 63–76. [CrossRef] [PubMed]
52. Delchier, N.; Ringling, C.; Cuvelier, M.-E.; Courtois, F.; Rychlik, M.; Renard, C.M. Thermal degradation of folates under varying oxygen conditions. *Food Chem.* **2014**, *165*, 85–91. [CrossRef] [PubMed]
53. European Medicines Agency. *ICH Topic Q 2 (R1): Validation of Analytical Procedures: Text and Methodology*; European Medicines Agency: London, UK, 2005.

54. United States Pharmacopoeial Convention. *United States Pharmacopoeia 42—National Formulary 37*; United States Pharmacopoeial Convention: Rockville, MD, USA, 2018.
55. Medicines and Healthcare Products Regulatory Agency. *British Pharmacopoeia 2019*; Medicines and Healthcare Products Regulatory Agency: London, UK, 2019.

Sample Availability: Samples of the proposed FDCs are available from the authors.

 © 2019 by the authors. Licensee MDPI, Basel, Switzerland. This article is an open access article distributed under the terms and conditions of the Creative Commons Attribution (CC BY) license (http://creativecommons.org/licenses/by/4.0/).

Article

Automated Stopped-Flow Fluorimetric Sensor for Biologically Active Adamantane Derivatives Based on Zone Fluidics

Paraskevas D. Tzanavaras [1,*], Sofia Papadimitriou [1] and Constantinos K. Zacharis [2]

1. Laboratory of Analytical Chemistry, School of Chemistry, Faculty of Sciences, Aristotle University of Thessaloniki, Thessaloniki GR-54124, Greece; paristzanavaras@gmail.com
2. Laboratory of Pharmaceutical Analysis, Department of Pharmaceutical Technology, School of Pharmacy, Aristotle University of Thessaloniki, Thessaloniki GR-54124, Greece; czacharis@pharm.auth.gr
* Correspondence: ptzanava@chem.auth.gr; Tel.: +30-2310997721; Fax: +30-2310997719

Received: 29 September 2019; Accepted: 1 November 2019; Published: 3 November 2019

Abstract: A zone-fluidics (ZF) based automated fluorimetric sensor for the determination of pharmaceutically active adamantine derivatives, i.e., amantadine (AMA), memantine (MEM) and rimantadine (RIM) is reported. Discrete zones of the analytes and reagents (o-phthalaldehyde and N-acetylcysteine) mix and react under stopped-flow conditions to yield fluorescent *iso*-indole derivatives ($\lambda_{ex}/\lambda_{em}$ = 340/455 nm). The proposed ZF sensor was developed and validated to prove suitable for quality control tests (assay and content uniformity) of commercially available formulations purchased from the Greek market (EU licensed) and from non-EU web-pharmacies at a sampling rate of 16 h^{-1}. Interestingly, a formulation obtained through the internet and produced in a third—non-EU—country (AMA capsules, 100 mg per cap), was found to be out of specifications (mean assay of 85.3%); a validated HPLC method was also applied for confirmatory purposes.

Keywords: memantine; rimantadine; amantadine; zone fluidics; o-phthalaldehyde; derivatization; stopped-flow; quality control

1. Introduction

Adamantane analogues namely amantadine (AMA) and rimantadine (RIM) have been extensively used for the treatment of influenza A virus infection for more than three decades [1,2]. The mechanism of their action is based on the inhibition of influenza A virus reproduction by hindering the M2 protein ion channel and thus preventing the release of viral RNA into the cytoplasm of infected cells [3,4]. On the other hand, another structurally analogous compound, memantine (MEM), is a noncompetitive N-methyl-D-aspartate receptor antagonist is widely used for the treatment of Alzheimer's and Parkinson's diseases, dementia syndromes and also more recently for the possible treatment of glaucoma [5,6].

Quality control is an important step during and after the production process of pharmaceutical formulations to ensure the safety and therapeutic efficacy of the final products. Additionally, due to the evolution of WEB drug suppliers and the existence of fake products in the market, the QC of commercially available pharmaceuticals is even more critical. On this basis, automation of the analytical process offers significant advantages in terms of rapidity and effectiveness [7]. The concept of Zone Fluidics (ZF) [8] combines some interesting features that are more than welcome in pharmaceutical analysis: (i) automation of chemical reactions through zones, (ii) single line configurations, (iii) generation of waste at the level of microliters, (iv) no need for reconfiguration in order to apply different methods, (v) robustness and unattended long term operation, etc.

A literature search (Scopus) revealed only a few automated flow methods reporting the determination of adamantine derivatives [9–13]. Most of them are based on potentiometric detection through suitable electrodes [9–11]. Although such approaches are interesting from an academic point of view, when it comes to real world QC industrial applications, the main disadvantage is the non-commercial availability of the electrodes. Other unattractive features include the fouling of the electrode surface in real sample analysis, the necessity for periodic activation/conditioning of the electrode [9], logarithmic regression equations, while only in the memantine potentiometric method proposed by Nashar et al., the authors presented validation results according to the international guidelines (i.e., robustness, ruggedness, etc.) [11]. On the other hand, in the chemiluminescence-based method reported by Alarfaj and El-Tohamy, the authors selected only the pH to validate the robustness of the SI procedure when, from the schematic effects of the other parameters, it is obvious that small variations in the flow rate, the aspiration volumes or the concentration of the reagents cause dramatic variations in the CL responses [12]. Finally, a more recent fluorimetric method proposed for the determination of amantadine is based on a competitive assay between the drug and the dye Thionine for the Cucurbit [8], uril cavity [13]. Although the authors claim adequate analytical features for application in both pharmaceuticals and urine samples, no proper validation experiments were presented to support the effectiveness of the procedure for QC purposes (i.e., robustness, determination range, residuals, etc.).

From an analytical chemistry point of view, the bottleneck for the development of simple methods for the direct quantitation of this class of compounds is the lack of chromophore or fluorophore moieties in their chemical structure. Derivatization through the primary amino group is therefore a viable solution in order to enhance their detectability. Among the various reagents that exist for the derivatization of primary amines [14], in this study, we selected o-phthalaldehyde (OPA) for the following reasons: (i) it reacts fast and under mild conditions, (ii) it is commercially available at low/reasonable cost, (iii) it is stable and simple to handle, (iv) it typically offers adequately sensitivity for most applications; on the other hand—compared to other reagents—OPA derivatives are less stable, but this is not a disadvantage when the reactions and detection are carried out on-line in a strictly reproducible analytical cycle [15].

The goals of the present study where, therefore, the following: (i) to examine the reaction of OPA with the three pharmaceutically active adamantine derivatives under flow conditions; (ii) to combine the advantages of OPA and the ZF automated technique; (iii) to develop and validate a simple and efficient platform for the analysis of pharmaceuticals from various sources in order to prove their safety and therapeutic efficacy.

2. Results and Discussion

2.1. Reactivity of MEM, AMA, RIM with o-Phthalaldehyde

Preliminary experiments confirmed that the analytes can react with OPA through their primary amino group under flow conditions, in the presence of a suitable nucleophilic agent (NAC) (Figure S2 in the supplementary material). However, it was interesting to notice that despite the structure similarities of the analytes, RIM showed a considerably higher reactivity in both terms of reaction speed and sensitivity, due to the fact that the primary amino group is attached to a ternary carbon atom. The order of reactivity (RIM >> AMA $\cong$ MEM) is in accordance with previous reports on the behavior of primary aliphatic amines and is affected mainly by steric hindrance and polar phenomena [16,17].

In order to improve the sensitivity of the analytical procedure, preliminary experiments were also focused on a stopped-flow step prior to FL detection. The advantage of the stopped-flow technique is the minimization of the radial dispersion effect that results in broad peaks and competes to the sensitivity gain due to the increased reaction time. Stopped-flow can be either applied in the flow cell or in a suitable coil before detection. The former approach is advantageous on the basis of monitoring the reaction in real time. However, our experiments proved that the OPA derivatives were not stable

2.2. Development of the ZF Sensor

The development of the ZF sensor involved experimental studies of the various chemical and instrumental parameters that are expected to affect the performance of the procedure. The investigated parameters, the studied range and the finally selected values are tabulated in Table 1. The selection criteria were a compromise among sensitivity, sampling rate and reagent consumption. The starting values of the variables during experiments were: $c(OPA) = 1.0$ mmol L^{-1}, $c(NAC) = 1.0$ mmol L^{-1} in pH = 10.0 (0.1 mol L^{-1} carbonate buffer), $\gamma(ATD) = \gamma(MEM) = 100$ mg L^{-1}, $\gamma(RIM) = 10$ mg L^{-1}, $Q_V(C) = 0.6$ mL min^{-1}, $V(AMA) = V(MEM) = V(RIM) = V(OPA) = V(NAC) = 50$ µL. The order of zones mixing had no practical effect on the analysis, and the sequence described in Section 2.3 was adopted (NAC/Buffer–OPA–Sample).

The effect of the stopped flow time was examined in the range of 0–180 s. The signals increased non-linearly in the studied range, with the sensitivity increasing 2-fold in the first 60 s for all analytes. The latter value was selected making a compromise between sensitivity and sampling rate.

The pH proved to be a critical variable for all compounds. Its effect was evaluated in the range of 9.0–12.0, using 100 mmol L^{-1} carbonate buffer. As can be seen in the experimental results in Figure 1, a clear maximum at a pH value of 10 was observed for all analytes. The type of buffer proved to have no effect (carbonate vs borate vs Britton–Robinson).

Table 1. Study of instrumental and chemical variables.

Variable	Studied Range	Selected Value
Instrumental		
$t(SF)/s$	0–180	60
$V(S)/\mu L$	50–100	100
$V(R)/\mu L$	25–100	75
$V(OPA)/\mu L$	25–100	50
Chemical		
pH	9.0–12.0	10.0
$c(NAC)/$mmol L^{-1}	0.5–5.0	2.5
$c(OPA)/$mmol L^{-1}	0.5–10.0	5.0

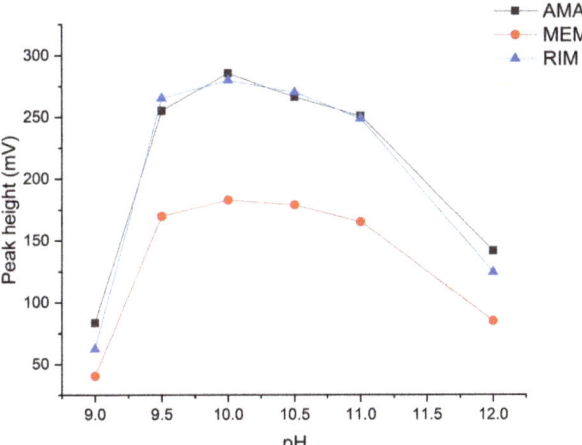

Figure 1. Effect of the pH on the derivatization reaction; for experimental details see Section 2.2. AMA = amantadine, RIM = rimantadine, MEM = memantine.

2.3. Validation of the ZF Sensor

The proposed sensor has been validated for determination range, limits of detection (LOD) and quantification (LOQ), precision, selectivity, accuracy and robustness.

The effects of the volume and amount concentration of the OPA and NAC reagents were investigated within the ranges included in Table 1. The criteria for selecting the final values were sensitivity and consumption. In all cases non-linear relationships were observed between the fluorescence intensities and the examined variables. The values of 75 µL (NAC solution, 2.5 mmol L^{-1}) and 50 µL (OPA solution, 5 mmol L^{-1}) provided adequate sensitivity and excess of reagents for the selected applications.

The effect of the sample injection volume is always an important parameter in flow-based sensors since it directly affects the dispersion of the analytes within the flow lines. The sensitivity increased between 150% and 200% for all analytes within 50–150 µL sample injection volume. Finally, the volume of 100 µL was selected as a compromise between sensitivity, consumption and sampling rate.

2.3. Validation of the ZF Sensor

The proposed sensor has been validated for determination range, limits of detection (LOD) and quantification (LOQ), precision, selectivity, accuracy and robustness.

The determination range was evaluated within 50% to 150% of the 100% level for the three analytes, i.e., 50–150 mg L^{-1} for AMA and MEM and 5–15 mg L^{-1} for RIM. The selected range covers the specifications for both assay and content uniformity tests. Six calibration levels were used ($n = 6$). The respective regression equations were

$$F = 72.9\ (\pm 5.8) + 4.84\ (\pm 0.05) \times \gamma(\text{AMA}),\ r^2 = 0.9996$$

$$F = 100.8\ (\pm 3.0) + 2.56\ (\pm 0.03) \times \gamma(\text{MEM}),\ r^2 = 0.9996$$

$$F = 17.0\ (\pm 6.1) + 42.28\ (\pm 0.95) \times \gamma(\text{RIM}),\ r^2 = 0.9990$$

where F is the fluorescence intensity as measured by the detector and γ the mass concentration of the analytes in mg L^{-1}. Linearity was further validated using the back-calculated concentrations (residuals). The percent residuals were distributed randomly around the "zero axis" and ranged < ± 3% (−1.5% to +0.9% for AMA, −0.2% to +1.1% for MEM and −1.6% to +1.7% for RIM).

The LODs and LOQs were estimated based on the following equations

$$\text{LOD} = 3.3 \times \text{SD}_b/s \text{ and } \text{LOQ} = 10 \times \text{SD}_b/s$$

where SD$_b$ is the standard deviation of the intercept and s is the slope of the respective regression lines. The calculated LOD/LOQ for the three analytes were 4/12 mg L^{-1} for AMA, 3.9/11.2 mg L^{-1} for MEM and 0.5/1.5 mg L^{-1} for RIM respectively.

The within-day precision was validated at the 100% level for the three active pharmaceutical ingredients, i.e., 100 mg L^{-1} for AMA and MEM and 10 mg L^{-1} for RIM. The RSD values were better than 1.9% for 8 consecutive injections. The day-to-day precision was validated by comparing the slopes of six regression lines for each analyte within a time period of 10 days. The RSD values of the slopes were better than 6.5% in all cases.

The selectivity and accuracy of the proposed sensor were validated using the "placebo approach" by preparing synthetic samples at three concentration levels of 50%, 100% and 150% for each compound. The synthetic samples also contained the placebo mixture (all excipients except for the active ingredients) at 1000 mg L^{-1} for AMA and MEM and 100 mg L^{-1} for RIM (10-fold excess over the 100% level). The acceptable recoveries should be in the range of 95% to 105%. The experimental results included in Table 2 confirmed the accuracy and selectivity of the procedure.

The robustness of the proposed method was evaluated by small deliberate variations (±5%) on critical instrumental and chemical parameters such as the stopped flow time (57–63 s), the sample volume (95–105 µL), the reaction pH (9.5–10.5), the amount concentration of OPA (4.7–5.3 mmol L^{-1}) and the amount concentration of NAC (2.3–2.7 mmol L^{-1}). With the exception of the pH all examined parameters resulted in variations in the range of 95% to 105% for all analytes. On the other hand, the strict regulation of the pH proved to be the most critical prerequisite since the variations ranged between 91.5% to 110.8%.

Table 2. Accuracy and selectivity of the proposed ZF method.

Sample	Analyte Concentration (mg L^{-1})	Placebo Concentration (mg L^{-1})	Recovery (%)	
			AMA	MEM
S1	50	1000	103.6	97.2
S2	100	1000	101.0	99.7
S3	150	1000	102.9	101.7
				RIM
S1	5	100		103.6
S2	10	100		97.1
S3	15	100		101.5

2.4. Applications of the ZF Sensor

The developed ZF fluorimetric sensor has been applied to the quality control (QC) of commercially available AMA and MEM containing pharmaceuticals (there were no commercially available RIM formulations in the Greek market). Two of the formulations, i.e., MEM oral drops (10 mg mL^{-1}) and AMA caps (100 mg per cap) were EU-licensed obtained from the Greek market, while two additional formulations, i.e., MEM tabs (5 mg per tab) and AMA(2) caps (100 mg per tab) were obtained from web-pharmacies and are produced in third countries (non-EU licensed). MEM oral drops were analyzed in terms of assay of the active ingredient, while capsules and tablets were also analyzed for content uniformity. The experimental results are presented in Tables 3 and 4. The Tables also include the specification limits of the pharmacopoeias for each formulation and the results from a corroborative, in-house validated HPLC method based on reversed phase separation and post-column derivatization (see Section 3.5 for details). The comparison of the obtained results (*t*-test) indicated that there is no statistical difference between the proposed ZF and HPLC methods at confidence interval 95% ($t_{exp} < t_{crit\ (95\%,\ n\ =\ 4)}$). The content uniformity results are also presented graphically, including the specification limits according to USP monograph (Figure 2).

As can be seen in the experimental results of Table 3, both EU-licensed formulations and the non-EU MEM tablets were within specifications regarding to the assay test. On the other hand, the non-EU licensed AMA capsules were out of specifications in the assay test with an average content of 85.3 (±2.1) mg AMA per capsule that is less than the 90% to 110% limits set by international pharmacopoeias. These results were found to be in accordance with the corroborative HPLC method. As can be seen in Table 4, only the EU-licensed AMA capsule formulation passed the content uniformity test based on the USP acceptance value (AV). Non-EU produced AMA and MEM formulations failed to pass the content uniformity test using both 10 and 30 units for analysis.

Table 3. Assay of AMA- and MEM-containing pharmaceutical formulations.

Formulation/Sample	Label Content	Specifications	Assay (%)	HPLC (%)
AMA caps (EU)	100 mg/cap	95% to 105%	97.4 (±1.5 [a])	96.1 (±2.2)
AMA caps (non-EU)	100 mg/cap	95% to 105%	85.3 (±1.2)	86.9 (±1.8)
MEM tabs (non-EU)	5 mg/tab	95% to 105%	92.4 (±1.7)	92.2 (±1.9)
MEM oral drops (EU)				
Lot1	10 mg mL^{-1}	95% to 105%	99.3 (±0.8)	100.5 (±1.7)
Lot2	10 mg mL^{-1}	95% to 105%	99.5 (±1.2)	101.2 (±2.6)
Lot3	10 mg mL^{-1}	95% to 105%	98.5 (±1.1)	99.8 (±2.3)

[a] Standard deviation.

Table 4. Content/Dosage uniformity test of pharmaceutical preparations.[a]

Sample		Content/Dosage		
MEM oral drops (EU)				
Lot1	10 mg mL⁻¹	95% to 105%	99.3 (±0.8)	100.5 (±1.7)
Lot2	10 mg mL⁻¹	95% to 105%	99.5 (±1.2)	101.2 (±2.6)
Lot3	10 mg mL⁻¹	95% to 105%	98.5 (±1.1)	99.8 (±2.3)

[a] standard deviation

Table 4. Content/Dosage uniformity test of pharmaceutical preparations.

Content uniformity (%)

Sample	AMA Caps (EU) 100 mg/Cap	AMA Caps (Non-EU) 100 mg/Cap	MEM Tabs (Non-EU) 5 mg/Tab
S1	92.1 (±1.2)	85.4 (±0.6)	79.8 (±1.5)
S2	98.3 (±0.5)	84.4 (±0.8)	85.5 (±1.7)
S3	98.9 (±0.5)	88.2 (±1.5)	100.8 (±2.1)
S4	103.7 (±0.9)	86.7 (±1.0)	93.5 (±2.2)
S5	101.3 (±1.8)	85.7 (±1.2)	82.5 (±1.7)
S6	96.0 (±1.1)	86.0 (±1.2)	100.8 (±1.4)
S7	96.3 (±0.9)	84.5 (±0.6)	96.2 (±0.9)
S8	94.9 (±1.3)	84.5 (±0.9)	86.6 (±2.5)
S9	94.2 (±1.4)	83.7 (±1.5)	109.7 (±1.2)
S10	104.2 (±0.8)	83.9 (±1.7)	88.6 (±2.1)
S11		85.0 (±1.1)	92.5 (±1.1)
S12		84.1 (±1.2)	96.7 (±1.3)
S13		85.6 (±0.3)	89.9 (±0.6)
S14		83.9 (±0.5)	87.6 (±0.8)
S15		83.6 (±0.9)	96.2 (±1.3)
S16		85.9 (±0.5)	88.4 (±0.2)
S17		86.2 (±1.2)	100.1 (±0.9)
S18		84.9 (±1.3)	87.5 (±1.6)
S19		84.8 (±0.8)	90.9 (±1.5)
S20		86.7 (±1.2)	106.1 (±1.6)
S21		82.9 (±1.6)	88.2 (±0.9)
S22		85.3 (±0.6)	88.9 (±0.8)
S23		85.6 (±0.8)	90.4 (±1.0)
S24		85.1 (±0.5)	105.8 (±1.3)
S25		84.7 (±1.0)	89.5 (±1.9)
S26		85.6 (±1.2)	91.8 (±1.8)
S27		86.0 (±0.9)	91.9 (±2.2)
S28		84.8 (±1.7)	105.7 (±2.4)
S29		83.9 (±1.1)	90.5 (±1.8)
S30		85.4 (±1.0)	85.4 (±0.7)
X̄	97.4	85.1	92.9
M	98.5	98.5	98.5
s	3.9	1.1	7.4
AV	10.5	15.6	20.5
L1	15	15	15
Result	Pass (AV<L1)	Fail (AV>L1)	Fail (AV>L1)

[a] Standard deviation.

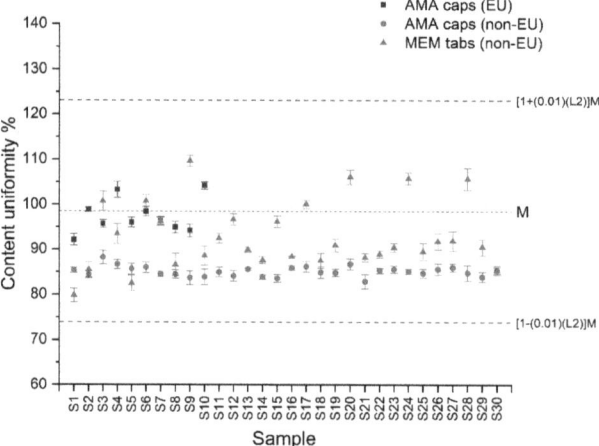

Figure 2. Graphical depiction of the content uniformity results of the selected EU and non-EU formulations using the proposed ZF method; LSL = lower specification limit, USL = upper specification limit.

3. Materials and Methods

3.1. Instrumentation

The home-made ZF setup (Figure 3) consisted of the following parts: a Minipuls3 peristaltic pump (Gilson), a micro-electrically actuated 10-port valve (Valco), a RF-551 flow-through spectrofluorimetric detector (Shimadzu); PTFE tubing was used for the connections of the flow

3. Materials and Methods

3.1. Instrumentation

The home-made ZF setup (Figure 3) consisted of the following parts: a Minipuls3 peristaltic pump (Gilson), a micro-electrically actuated 10-port valve (Valco), a RF-551 flow-through spectrofluorimetric detector (Shimadzu); PTFE tubing was used for the connections of the flow configuration (0.5 or 0.7 mm i.d.), and Tygon tubing was used in the peristaltic pump. Control of the ZF system was performed through a LabVIEW (National Instruments) based program also developed in house.

The HPLC-PCD setup comprised the following parts (Figure S1 in supplementary material): a AS3000 autosampler (Thermo Scientific), a LC-9A binary pump (Shimadzu, Japan), a RF-551 spectrofluorimetric detector ($\lambda_{ex}/\lambda_{em}$ = 340/455 nm) (Shimadzu, Japan), an Elite^{TM} vacuum degasser (Alltech, U.S.), a 150 × 4.6 mm i.d. reversed-phase column (Prevail, Alltech). A peristaltic pump was employed for the propulsion of the PCD reagents (Gilson Minipuls3). PTFE tubing (0.5 mm i.d.) was used for the construction of the reaction coil (RC) and all necessary PCD connections.

Data acquisition including peak highs (ZF method) and areas (HPLC-PCD method) was carried out through the Clarity® software (version 4.0.3, DataApex, Czech Republic).

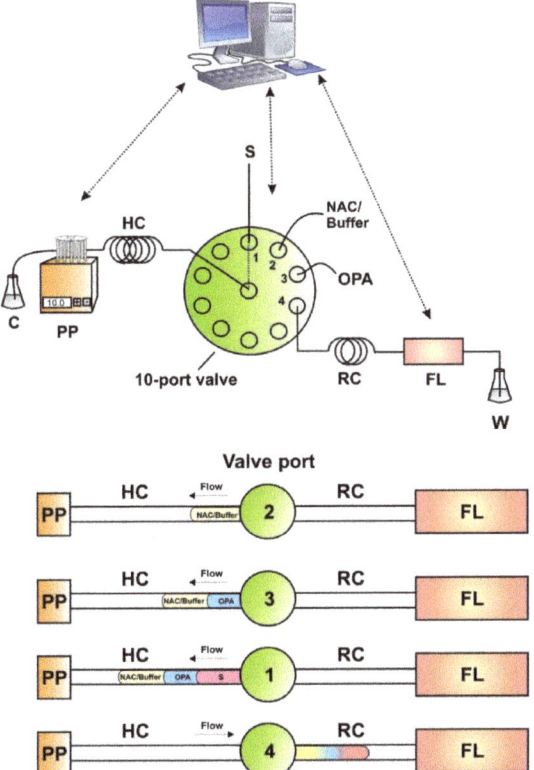

Figure 3. Graphical depiction of the Zone Fluidics manifold and the analytical sequence steps for the determination of the adamantane derivatives; PP = peristaltic pump, HC = holding coil, S = sample, FL = fluorimetric detector, RC = reaction coil, C = carrier (water), W = waste, OPA = o-phthalaldehyde, NAC = N-acetylcysteine.

3.2. Reagents and Solutions

Memantine (MEM), Amantadine (AMA) and Rimantadine (RIM) were purchased by Sigma; o-phthalaldehyde (OPA, Fluka), N-acetylcysteine (NAC, Merck), Na$_2$CO$_3$ (Merck), NaOH (Merck) and HCl (Sigma) were all of analytical grade. Doubly de-ionized water was produced by a Milli-Q system (Millipore).

3.2. Reagents and Solutions

Memantine (MEM), Amantadine (AMA) and Rimantadine (RIM) were purchased by Sigma; *o*-phthalaldehyde (OPA, Fluka), *N*-acetylcysteine (NAC, Merck), Na_2CO_3 (Merck), NaOH (Merck) and HCl (Sigma) were all of analytical grade. Doubly de-ionized water was produced by a Milli-Q system (Millipore).

The stock solutions of the analytes were prepared at the 1000 mg L^{-1} level in water. They were kept refrigerated and were found to be stable for at least one month. Working solutions were prepared on a daily basis by appropriate dilutions in water. The 100% level was 100 mg L^{-1} for MEM and AMA and 10 mg L^{-1} for RIM.

The derivatizing reagent (OPA) was prepared at an amount concentration of 10 mmol L^{-1} by firstly dissolving in 0.5 mL methanol and subsequently adding 9.5 mL water [19]. This solution was stable for two weeks at 4 °C in an aluminum foil-wrapped container. NAC was prepared in water (10 mmol L^{-1}) and diluted at the desired working concentrations in 100 mmol L^{-1} carbonate buffer (pH = 10).

The placebo mixture used in accuracy and selectivity studies consisted of representative pharmaceutical grade excipients and was prepared at a nominal concentration of 10 mg mL^{-1} according to previously published protocols [20].

3.3. ZF Procedure

A graphical depiction of the ZF sequence can be found in Figure 3. In brief, NAC/Buffer (75 µL, 2.5 mmol L^{-1}, pH = 10), OPA (50 µL, 5 mmol L^{-1}) and sample (100 µL) were sequentially aspirated in the holding coil (HC). The reaction mixture was propelled for 30 s towards the reaction coil (RC) at a flow rate of 0.6 mL min^{-1}, and the reaction was allowed to develop for 60 s under stopped-flow conditions. Detection was carried out downstream at $\lambda_{ex}/\lambda_{em}$ = 340/455 nm. More detailed description of the ZF steps is tabulated in Table S1 (supplementary material). The sampling throughput was 16 h^{-1}.

3.4. Preparation of Samples

Pharmaceutical samples included both EU-licensed products—purchased from local pharmacies—and non-EU-licensed formulations—purchased from web-pharmacies. Three lots of MEM oral drops (10 mg mL^{-1}) were analyzed directly, after 10-fold dilution in water. AMA capsules (100 mg per cap) and MEM tablets (5 mg per tab) (n = 10) were ultrasonically dissolved in water followed by filtration through 0.45 µm syringe filters and dilution with water to meet the 100% level concentration. For content uniformity tests, ten capsules or tablets from each formulation were treated individually (n = 10) as indicated by the U.S. pharmacopoeia [21].

3.5. HPLC-PCD Corroborative Method

Fifty microliters of samples or standards were injected in HPLC column and separated at ambient temperature at a flow rate of 0.5 mL min^{-1}. The HPLC mobile phase was CH_3OH/phosphate buffer (25 mmol L^{-1}, pH = 2.0) at a volume ratio of 50:50. Prior to use, it was filtered under vacuum through 0.22 µm membrane filters (Whatman®). The post-column-derivatization reagents were OPA at an amount concentration of 20 mmol L^{-1} and a mixture of NAC (5 mmol L^{-1})/100 mmol L^{-1} borate buffer (pH = 11.0). The PCD reagents were pre-mixed on-line (0.2 mL min^{-1} each) through a binary inlet static mixer (BISM, ASI-Analytical Scientific Instruments) with internal volume of 250 µL. The derivatization was allowed to proceed on passage through a 100 cm long knitted reaction coil. The derivatives of Amantadine, (t_R = 6.6 min), Rimantadine (t_R = 13.5 min), Memantine (t_R = 15.8 min) were detected fluorimetrically at $\lambda_{ex}/\lambda_{em}$ = 340/455 nm (see typical chromatogram in Figure S3 in the supplementary material).

4. Conclusions

The developed automated fluorimetric sensor for the analysis of pharmaceutically active adamantane derivatives offers some interesting features:

1. This is the first flow-based method for this class of active pharmaceutical ingredients;
2. The sensor utilizes commercially available reagents, is simple and straightforward;
3. The concept of zone fluidics minimizes the consumption of the samples and reagents compared to continuous flow methods, at a sampling rate of 16 h^{-1};
4. Validation experiments confirmed the suitability of the method for the QC of commercially available formulations;
5. A non-EU licensed formulation was found to be out of the specifications set by international pharmacopoeias;

Supplementary Materials: The following are available online at http://www.mdpi.com/1420-3049/24/21/3975/s1, Figure S1: Schematic diagram of the HPLC-PCD setup; Figure S2: Reaction between OPA and the studied adamantane derivatives using N-acetylcysteine (NAC) as nucleophilic reagent, Figure S3: Representative chromatogram from standard mixture of the adamantane derivatives by the corroborative HPLC-PCD method, Table S1: ZF sequence for the automated determination of adamantane derivatives.

Author Contributions: Conceptualization, P.D.T. and C.K.Z.; methodology, P.D.T. and C.K.Z.; validation, P.D.T., S.P. and C.K.Z.; data curation, P.D.T., S.P. and C.K.Z; writing—original draft preparation, P.D.T. and C.K.Z.; writing—review and editing, P.D.T. and C.K.Z.

Funding: This research received no external funding.

Conflicts of Interest: The authors declare no conflict of interest.

References

1. Dolin, R.; Reichman, R.C.; Madore, H.P.; Maynard, R.; Linton, P.N.; Webber-Jones, J. A controlled trial of amantadine and rimantadine in the prophylaxis of influenza a infection. *N. Eng. J. Med.* **1982**, *307*, 580–584. [CrossRef] [PubMed]
2. Tominack, R.L.; Hayden, F.G. Rimantadine hydrochloride and amantadine hydrochloride use in influenza a virus infections. *Infect. Dis. Clin. N. Am.* **1987**, *1*, 459–478.
3. Wang, C.; Takeuchi, K.; Pinto, L.H.; Lamb, R. A ion channel activity of influenza a. virus M2 protein: Characterization of the amantadine block. *J. Virol.* **1993**, *67*, 5585–5594. [PubMed]
4. Takeda, M.; Pekosz, A.; Shuck, K.; Pinto, L.H.; Lamb, R.A. Influenza a virus m2 ion channel activity is essential for efficient replication in tissue culture. *J. Virol.* **2002**, *76*, 1391–1399. [CrossRef] [PubMed]
5. Schneider, E.; Fischer, P.A.; Clemens, R.; Balzereit, F.; Fünfgeld, E.W.; Haase, H.J. Effects of oral memantine administration on Parkinson symptoms. Results of a placebo-controlled multicenter study. *Dtsch. Med. Wochenschr.* **1984**, *109*, 987–990. [CrossRef] [PubMed]
6. Yücel, Y.H. Memantine protects neurons from shrinkage in the lateral geniculate nucleus in experimental glaucoma. *Arch. Ophthalmol.* **2006**, *124*, 217–225. [CrossRef] [PubMed]
7. Marshall, G.; Wolcott, D.; Olson, D. Zone fluidics in flow analysis: Potentialities and applications. *Anal. Chim. Acta* **2003**, *499*, 29–40. [CrossRef]
8. Tzanavaras, P.D.; Themelis, D.G. Review of recent applications of flow injection spectrophotometry to pharmaceutical analysis. *Anal. Chim. Acta* **2007**, *588*, 1–9. [CrossRef] [PubMed]
9. Abdel-Ghani, N.T.; Shoukry, A.F.; Hussein, S.H. Flow injection potentiometric determination of amantadine HCl. *J. Pharm. Biomed. Anal.* **2002**, *30*, 601–611. [CrossRef]
10. Amorim, C.G.; Araújo, A.N.; Montenegro, M.C.B.S.M.; Silva, V.L. Sequential injection lab-on-valve procedure for the determination of amantadine using potentiometric methods. *Electroanalysis* **2007**, *19*, 2227–2233. [CrossRef]
11. El Nashar, R.M.; El-tantawy, A.S.M.; Hassan, S.S.M. Potentiometric membrane sensors for the selective determination of memantine hydrochloride in pharmaceutical preparations. *Int. J. Electrochem. Sci.* **2012**, *7*, 10802–10817.

12. Alarfaj, N.A.; El-Tohamy, M.F. A sensitive sequential injection analysis (SIA) determination of memantine hydrochloride using luminol-hydrogen peroxide induced chemiluminescence detection. *J. Chil. Chem. Soc.* **2015**, *59*, 2657–2661. [CrossRef]
13. Del Pozo, M.; Fernández, Á.; Quintana, C. On-line competitive host-guest interactions in a turn-on fluorometric method to amantadine determination in human serum and pharmaceutical formulations. *Talanta* **2018**, *179*, 124–130. [CrossRef] [PubMed]
14. Rigas, P.G. Review: Liquid chromatography-post-column derivatization for amino acid analysis: Strategies, instrumentation, and applications. *Instrum. Sci. Technol.* **2012**, *40*, 161–193. [CrossRef]
15. Zacharis, C.K.; Tzanavaras, P.D. Liquid chromatography coupled to on-line post column derivatization for the determination of organic compounds: A review on instrumentation and chemistries. *Anal. Chim. Acta* **2013**, *798*, 1–24. [CrossRef] [PubMed]
16. Hill, R.L.; Crowell, T.I. Structural Effects in the Reactivity of Primary Amines with Piperonal. *J. Am. Chem. Soc.* **1956**, *78*, 2284–2286. [CrossRef]
17. Murthy, C.P.; Sethuram, B.; Navaneeth Rao, T. Kinetics and mechanism of oxidation of some aliphatic amines by ditelluratocuprate (III). *Mon. für Chem. Chem. Mon.* **1982**, *113*, 941–948. [CrossRef]
18. Hanczkó, R.; Jámbor, A.; Perl, A.; Molnár-Perl, I. Advances in the o-phthalaldehyde derivatizations. *J. Chromatogr. A* **2007**, *1163*, 25–42. [CrossRef] [PubMed]
19. Genfa, Z.; Dasgupta, P.K. Fluorometric measurement of aqueous ammonium ion in a flow injection system. *Anal. Chem.* **1989**, *61*, 408–412. [CrossRef]
20. Tzanavaras, P.D.; Karakosta, T.D. Automated tagging of pharmaceutically active thiols under flow conditions using monobromobimane. *J. Pharm. Biomed. Anal.* **2011**, *54*, 882–885. [CrossRef] [PubMed]
21. USP-NF <905> General Chapter, Uniformity of Dosage unit. Available online: https://www.usp.org (accessed on 1 November 2019).

Sample Availability: Samples of the compounds mentioned in the text are not available from the authors but they can be purchased from the manufacturers mentioned in the "reagents and solutions" section.

© 2019 by the authors. Licensee MDPI, Basel, Switzerland. This article is an open access article distributed under the terms and conditions of the Creative Commons Attribution (CC BY) license (http://creativecommons.org/licenses/by/4.0/).

Article

Determination of Polyhexamethylene Biguanide Hydrochloride Using a Lactone-Rhodamine B-Based Fluorescence Optode

Akane Funaki, Yuta Horikoshi, Teruyuki Kobayashi and Takashi Masadome *

Department of Applied Chemistry, Faculty of Engineering, Shibaura Institute of Technology, Toyosu, Koto-ku, Tokyo 135-8548, Japan; d09077@shibaura-it.ac.jp (A.F.); m310025@shibaura-it.ac.jp (Y.H.); d08043@shibaura-it.ac.jp (T.K.)
* Correspondence: masadome@sic.shibaura-it.ac.jp

Received: 23 December 2019; Accepted: 7 January 2020; Published: 9 January 2020

Abstract: A new determination method for polyhexamethylene biguanide hydrochloride (PHMB) using a lactone-rhodamine B (L-RB) based fluorescence optode has been developed. The optode membrane consists of 2-nitrophenyl octyl ether as a plasticizer, L-RB, and poly (vinyl chloride). The optode responds to tetrakis (4-fluorophenyl) borate, sodium salt (NaTPBF) in the µM range. The fluorescence intensity of the L-RB film for PHMB solution containing 20 µM NaTPBF decreased linearly as the concentration of the PHMB solution increased in the concentration range from 0 to 8.0 µM, which shows that PHMB with a concentration range of 0 to 8.0 µM is determined by the L-RB film optode. The concentration of PHMB in the contact lens detergents by the proposed method was in accord with its nominal concentration.

Keywords: optode; polyhexamethylene biguanide hydrochloride; lactone-rhodamine B; contact-lens detergent

1. Introduction

Polyhexamethylene biguanide hydrochloride (PHMB), a kind of a cationic polyelectrolyte (CP), is very useful for disinfectants in a contact lens detergent (CLD) and sanitizers for swimming pools. Many analytical methods for PHMB have been studied [1–12]. However, the lower determination concentration of several analytical methods for PHMB is above a few ppm, and the methods have very tedious operation procedures and use very toxic reagents frequently. Furthermore, the methods were not applied to the measurement of PHMB in commercially available CLDs because the concentration of PHMB in the commercially available CLDs is ca. 1.1 ppm (5.0 µM). A highly sensitive HPLC method with a solid phase extraction and an evaporative light scattering detector, whose lower detection limit is 0.1 ppm (0.45 µM), has been reported for the determination of PHMB in CLDs [4]. However, the apparatus of the method is very expensive and exaggerated. On the other hand, we have reported a new spectrophotometric flow injection analysis (FIA) method for the determination of PHMB and determined PHMB in the CLDs by the FIA method [11]. However, a large amount of reagent solution was consumed. Recently, Uematsu et al. reported a new promising determination method of PHMB using a glucose oxidase enzymatic reaction. The method is relatively simple and has high sensitivity (determination range of 0.05 to 0.4 ppm). However, the method utilizes a relatively high-cost enzyme such as a glucose oxidase and catalase [12].

Therefore, a simpler and low-cost analytical method with a low amount of used reagent solution and waste for the measurement of PHMB in CLDs using a chemical sensor is very helpful. The chemical sensor is very useful analytical tool, since it has the advantages of simple operation and relatively low cost. A highly sensitive electroanalytical sensor for PHMB by using adsorptive voltammetry as reported by Hattori et al. did not apply to the determination of PHMB in CLDs [6]. We have already reported an FIA method for a CP and an anionic polyelectrolyte (AP), utilizing the formation of an ion associate between a cationic polyelectrolyte and AP or an anionic surfactant (AS), and an ion-selective electrode (ISE) detector [13,14].

An analytical method of PHMB using an optode as a chemical sensor is very promising, because the optodes have several excellent advantages that are distinct from the electrochemical measurements such as the adsorptive voltammetric electroanalytical sensor and the ISE, in that the optodes are less susceptible to electrical noise, do not require a lead wire from their sensing membrane, and do not require a reference electrode for measurement. However, no research papers regarding the determination method for PHMB using an optode have been published so far.

Recently, we have reported an AS-sensitive optode based on a lactone rhodamine B (L-RB) film [15–17]. From the results, PHMB is expected to be determined with an optode responding to a hydrophobic anion such as AS, because PHMB will form an ion associate with the hydrophobic anion, and the concentration of the free hydrophobic anion is detected by the AS optode.

In this work, we have developed the determination method of PHMB in CLDs using an optode based on L-RB film.

2. Results and Discussion

2.1. Absorption Spectra, Excitation, and Fluorescence Spectra of an L-RB Film after Immersing to a NaTPBF Solution

The sample solution contains PHMB and a hydrophobic anion, NaTPBF (constant concentration). An increase of concentration of PHMB decreases the concentration of free NaTPBF, because PHMB forms an ion associate with NaTPBF quantitatively. PHMB is determined by measuring the decrease of concentration of free NaTPBF by the optode based on L-RB film.

At first, we examined the response of the optode based on L-RB film to NaTPBF. Figure 1 illustrates the absorption spectra of the L-RB film after the L-RB film is immersed in NaTPBF solution containing a buffer solution (pH 4.0) for 30 min. The absorbance of the L-RB film at 559 nm shown in Figure 1 increased as the concentration of NaTPBF increases in the 0 to 10.0 μM region (r^2 = 0.980), as indicated in Figure 2. The graph equation for NaTPBF was 0.0081 x + 0.0296. Here, x is the μM concentration of NaTPBF, and y is the absorbance of the L-RB film. Next, the fluorescence of the L-RB film was measured in order to determine NaTPBF in the lower concentration region. Figure 3 depicts the excitation and fluorescence spectra of the L-RB film after it was immersing to a 5.0 μM NaTPBF solution (pH 4.0, adjusted with a 0.1 M CH_3COOH/CH_3COONa buffer solution) and the pH 4.0 buffer solution (0 μM NaTPBF solution) for 30 min, respectively. It can be seen that in case the sensing film is immersed in the NaTPBF solution, the fluorescence intensity of the sensing film significantly increased, compared with that of the case of immersion of the sensing film to pH 4.0 buffer solution. When the L-RB film was immersed in a NaTPBF solution, the maximum fluorescence intensity of the L-RB film was obtained at λ_{ex} = 561 nm and λ_{em} = 584 nm. In the subsequent experiments, the wavelengths were used for the fluorescence intensity measurements of the L-RB film. The increase of the fluorescence intensity at λ_{ex} = 561 nm and λ_{em} = 584 nm for the L-RB film immersed to the 5.0 μM NaTPBF solution was apparent due to a coextraction of the $TPBF^-$ anion and a proton into the L-RB film. Here, the L-RB protonates and associates with the extracted $TPBF^-$ anion; as a result, the protonation of the L-RB results in a spectral change. The formation reaction of the ion associate between the protonated L-RB and extracted $TPBF^-$ anion was as follows:

$$(TPBF^-)_{aq} + (H^+)_{aq} + (L\text{-}RB)_{memb} \text{ (low fluorescence intensity)}$$
$$\rightleftarrows (L\text{-}RB\text{-}H^+\text{-}TPBF^-)_{memb} \text{ (high fluorescence intensity)}$$

where memb is the L-RB film, and aq is an aqueous solution.

approach in which the fluorescence intensity of the optode is measured at a fixed time. An immersion time of 10 min was used in subsequent fluorescence experiments as a compromise of the sensitivity of the L-RB film and sample throughput. A good linear relationship ($r^2 = 0.980$) was also obtained between the fluorescence intensity at 561 nm and the NaTPBF concentration in the 0 to 10.0 μM range.

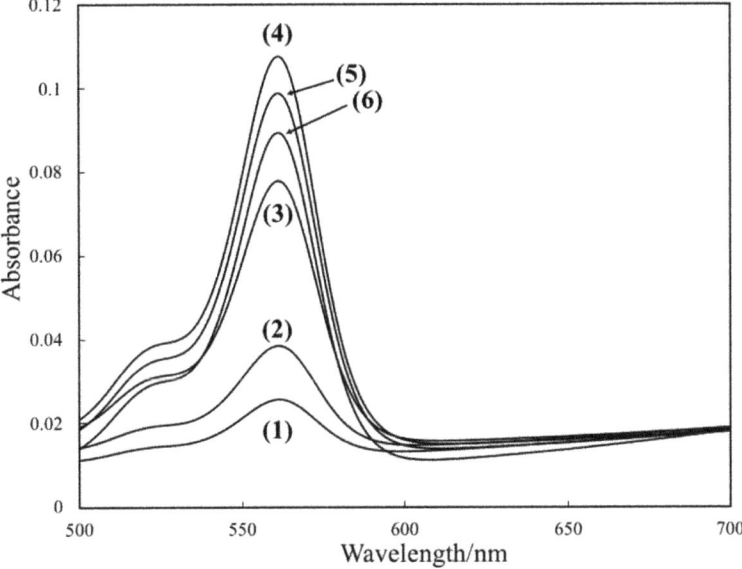

Figure 1. The absorption spectra of the lactone-rhodamine B (L-RB) film after it was immersed to a NaTPBF solution (pH 4.0, adjusted with a 0.1 M CH$_3$COOH/CH$_3$COONa buffer solution) for 30 min, respectively. Concentration of NaTPBF: (1) 0 μM, (2) 1.0 μM, (3) 5.0 μM, (4) 10.0 μM, (5) 50.0 μM, and (6) 100.0 μM.

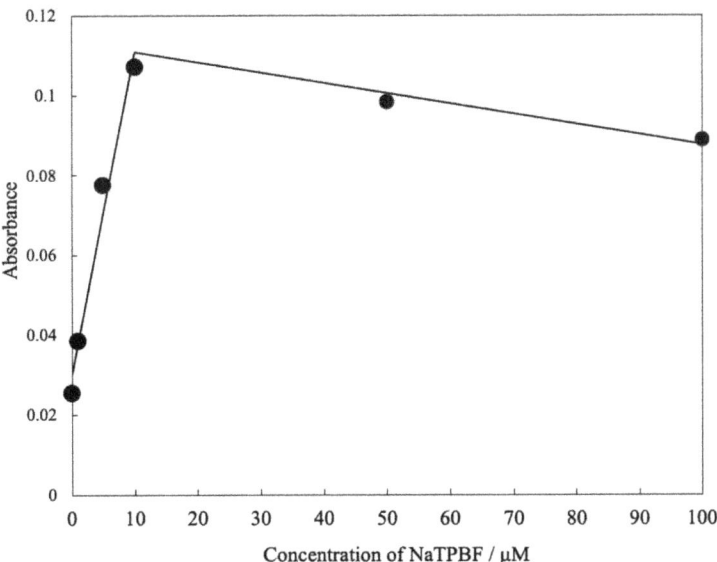

Figure 2. Calibration curve for NaTPBF obtained by using an L-RB-based optode.

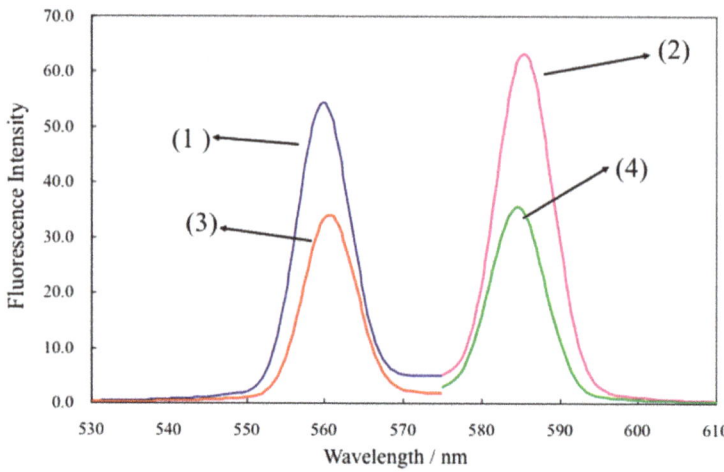

Figure 3. The excitation and fluorescence spectra of the L-RB film after it was immersing to a 5.0 µM NaTPBF solution (pH 4.0, adjusted with a 0.1 M CH₃COOH/CH₃COONa buffer solution) (1: excitation spectrum), (2: fluorescence spectrum), and that of the pH 4.0 buffer solution (0 µM NaDBS solution) (3: excitation spectrum), (4: fluorescence spectrum) for 30 min, respectively.

2.2. Calibration Curve for PHMB

The effect of a coexisting concentration of NaTPBF in a sample solution on the response sensitivity of the L-RB film to the PHMB anion was examined. If 100 µM was used as the coexisting concentration of NaTPBF, the sensitivity of the optode was much less than that when 10.0 and 20.0 µM are used as the coexisting concentrations of NaTPBF. In the event that 10.0 and 20.0 µM are used as coexisting concentrations of NaTPBF, the sensitivity of the optode is almost the same. When 10.0 µM is used as the coexisting concentration of NaTPBF, the concentration of PHMB in the commercially available contact lens detergents was not agreement with its nominal concentration in a preliminary examination (see Section 2.4). Since the determination concentration range of the optode for PHMB is enough to measure PHMB in the commercially available CLDs (ca. 1.1 ppm; 5.0 µM) when 20 µM is used for the coexisting concentration of NaTPBF, other concentrations of NaTPBF were not investigated. From the finding, the concentration of NaTPBF was determined to be 20.0 µM. Figure 4 reveals a standard calibration curve for PHMB in three measurements, which were obtained by using the L-RB film measured at λ$_{ex}$ = 561 nm and λ$_{em}$ = 584 nm after the immersion of the L-RB film to a PHMB solution in the presence of 20 µM NaTPBF for 10 min. The error bars show the average ± relative standard deviation. A plot of the fluorescence intensity of the L-RB film versus the concentration of PHMB ranged from 0 to 8.0 µM yields a straight line (r^2 = 0.989). The regression equation of the calibration curve for PHMB is $y = -13.0 x + 133$. Here, x is the µM concentration of PHMB, and y is the relative fluorescence intensity. The detection limit of the optode, which is defined as the concentration equivalent to 3σ, was ca. 1.6 µM, where σ is the standard deviation of the response for a blank solution.

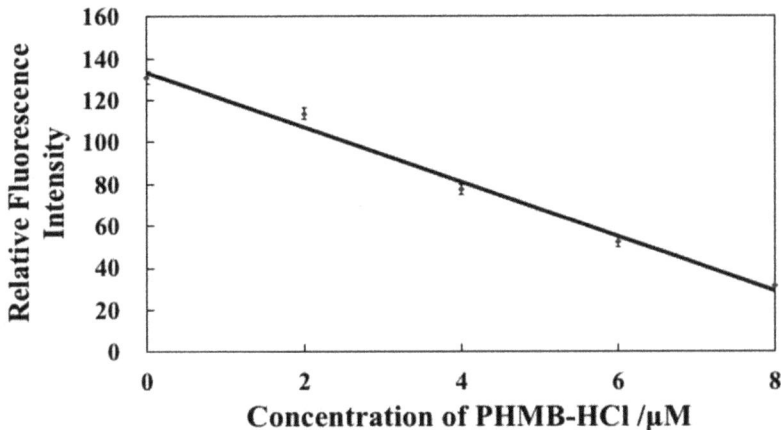

Figure 4. Calibration curve for polyhexamethylene biguanide hydrochloride (PHMB) in the presence of 20.0 μM NaTPBF obtained by using an L-RB-based fluorescence optode.

The tetrahydrofuran (THF) solution of the components of the L-RB film was stored for at least six months in a refrigerator at 4 °C while maintaining its performance as a sensing film for PHMB.

2.3. Evaluation of the Removal of PHMB by Using Several Solid-Phase Extraction Columns

Commercially available CLDs often contain polyhexanide (PHMB, 1.1 ppm, equivalent to 5.0 μM), buffer agent, stabilizer, an isotonic agent, pH adjuster, poloxamine (nonionic surfactant), and an ingredient called hydranate, which is known by chemists as hydroxyalkylphosphonate. In order to quantitate PHMB in CLDs by utilizing a standard addition method, the CLDs containing no PHMB must be prepared. Therefore, the removal of PHMB using several solid-phase extraction (SPE) columns was evaluated for 6.0 μM PHMB as a sample solution. Table 1 shows the sequence of adsorption of PHMB by solid-phase extraction columns. The sequence is as follows: Sep-Pak Plus CN > Sep-Pak Plus C18 > Sep-Pak Plus C8 > Sep-Pak Plus PS 2 > OASIS HLB 1cc. In case that Sep-Pak Plus CN as a solid-phase extraction column was used, approximately all the PHMB was removed. From the result, Sep-Pak Plus CN as a solid-phase extraction column was used for removal of PHMB in order to quantify PHMB in the CLDs by a standard addition method.

Table 1: Evaluation of the removal of PHMB by using several solid-phase extraction columns.

Solid-Phase Extraction Column	Concentration of Adsorbed PHMB Column	Removal of PHMB in from the Column/%
Sep-Pak Plus CN	6.0 μM	100
Sep-Pak Plus C18	5.0 μM	83
Sep-Pak Plus C8	2.8 μM	47
OASIS HLB 1cc	1.0 μM	17
Sep-Pak Plus PS 2	2.1 μM	35

The peak height for 6.0 μM PHMB with the SPE columns was compared with that without those columns, and the concentration of adsorbed PHMB in the columns was measured from the calibration curve for PHMB without the columns.

2.4. Determination of PHMB in the CLDs

The fluorescence intensities of the L-RB film for the solutions containing 20.0 μM of NaTPBF and a known amount of PHMB to a fivefold diluted CLD solution (A), and those for a solution containing 20.0 μM NaTPBF and a fivefold diluted CLD solution (A) through a Sep-Pak CN column were measured, respectively. The fluorescence intensity for a solution containing 20.0 μM NaTPBF and a a fivefold diluted CLD solution (A) through Sep-Pak C18 column was deducted from that for the solution containing 20 µM of PHMB to add to the fivefold diluted contact lens detergent solution without Sep-Pak C18 column. Figure 5 shows a plot of the deducted fluorescence intensity on the vertical axis and the PHMB concentration as the horizontal axis. From the Y-intercept of the straight line, the concentration of PHMB in the CLD (A) is calculated to be 5.1 µM, which is similar to the nominal PHMB concentration (5.0 µM) in Table 2. Table 2 shows the results for determination of PHMB in CLD (A) and (B) from Table 2, the RB film-based optode was capable of determining PHMB in a few CLDs.

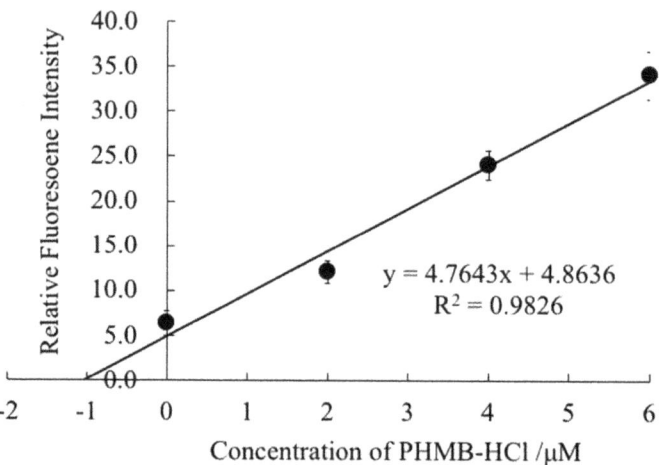

Figure 5. Determination of PHMB in a contact lens detergent (A) by the standard addition method. The fluorescence intensity for Sep-Pak C18 containing 20.0 µM of NaTPB and fivefold diluted contact lens detergent (A) through Sep-Pak C18 column was deducted that for the solutions containing 20.0 µM without Sep-Pak C18 column. The deducted PHMB concentration as the contact lens detergent solution for (A) added PHMB Sep-Pak C18 column through the deducted fluorescence intensity as the vertical axis was plotted for added PHMB concentration as the horizontal axis.

Table 2. Determination of PHMB in contact lens detergents.

Contact Lens Detergents	Nominal Value	Determination Value/µM
(A)	0.001 mg/mL (4.6 µM)	5.1
(B)	0.001 mg/mL (4.6 µM)	4.8

3. Experimental

3.1. Reagents

Poly(vinyl chloride) (PVC) powder and rhodamine B (RB) from Wako Pure Chemical Industries, Ltd. (Osaka, Japan were used). Sodium tetrakis(4-fluorophenyl)borate, Sep-Pak Plus C18, Sep-Pak Plus CN, Sep-Pak Plus CN Short, Sep-Pak Plus C18 Short, OASIS HLB were from Dojindo Sep-Pak Plus (Kumamoto, Japan) Sep-Pak PS (Sep-Pak Plus PS) column purchased from Waters (Milford, Massachusetts United States) reagents used were solid phase extraction (SPE). The PHMB concentration (mol/L) and PHMB indicates the number of moles of ionic groups/volume (liter). The of polymer solution concentration (M = mol dm⁻³) indicates the number of moles of ionic groups/volume (liter) of polymer solution.

3.2. Preparation of Optode Film Containing L-RB

3.2. Preparation of Optode Film Containing L-RB

An optode film containing L-RB was prepared as follows. The fabrication method is similar to our previous paper [14–16]. Take NPOE (5.0 g) and 24.0 mL of a mixed aqueous solution containing 75.0 µM RB and 1.0 M NaOH in a plastic tube, stir for 24 h at 55 °C, and extract L-RB into the NPOE phase. The NPOE phase (4.0 g) centrifuged from the aqueous phase (10 min) and PVC powder (0.80 g) were solubilized in 20 mL of tetrahydrofuran (THF). By means of immersing a quartz glass plate (length: 3.5 cm, width: 1.0 cm, thickness: 1 mm) in the THF solution for 5 s, an L-RB film for the measurement of fluorescence was fabricated. In the event of measurement of absorbance of the L-RB film, a glass plate (length of 3.5 cm, a width of 1.0 cm and thickness of 1 mm) was used in place of the quartz glass plate. The L-RB film was dried at room temperature for more than 2 h and then conditioned with CH_3COOH-CH_3COONa buffer solution (pH 4.0) for 24 h before starting an experiment. The fluorescence intensity of the L-RB film was measured for 10 min at λ_{ex} = 561 nm and λ_{em} = 584 nm on a spectrofluorimeter (JASCO FP-750, JASCO Corporation, Tokyo, Japan). The measurement was performed by placing the L-RB film on the diagonal of a 1 cm square quartz cuvette (height 4.5 cm, length 1.25 cm, width 1.25 cm) containing a sample solution (3.0 mL). The thickness of the prepared L-RB film was about 0.2 mm. In the case of measurement of absorbance of the L-RB film, glass plates coated with L-RB films were fixed in disposable cuvettes (4.5 cm high, 1.25 cm long, and 1.25 cm wide) filled with 2.0 mL of PHMB solution (pH 4.0 adjusted with a 0.1M CH_3COOH/CH_3COONa buffer solution). The absorption spectrum of the L-RB films was measured with a spectrophotometer (JASCO, V-530-iRM, JASCO Corporation, Tokyo, Japan). All spectroscopic measurements were performed in batch mode. The L-RB film was used as a disposable sensor for one-shot measurement.

3.3. Evaluation of Removal of PHMB by Using Several Solid-Phase Extraction Columns with a Flow Injection Analysis (FIA)

The FIA system and its experimental conditions for evaluation of the removal of PHMB in commercially available CLDs using several solid-phase extraction (SPE) columns are the same as that for the measurement of PHMB in the previous paper [11]. A PHMB sample solution (140 µL) with and without pretreatment using the SPE columns was injected into the FIA system. The peak height for 6.0 µM PHMB with the SPE columns was compared with that without those columns, and the concentration of adsorbed PHMB in the columns was measured from the calibration curve for PHMB without the columns.

4. Conclusions

A fluorescence optode based on an L-RB film for the determination of PHMB was fabricated. The response of the optode shows a good linear response between fluorescence intensity and the concentration of PHMB with the concentration range of 0 to 8.0 µM. The application of the L-RB film optode was made for the determination of PHMB in the CLDs. The procedures of the proposed method based on the L-RB film optode is much simpler than those of the conventional determination methods without using very toxic reagents. The L-RB film is easily fabricated with relatively low cost. The present method will be very promising for the quality control of CLDs. The L-RB film was used as a reversible film for the FIA of AS [16]. Therefore, the L-RB film will be used as a reversible film for the determination of PHMB. We continue the research regarding the FIA of PHMB using the optode detector based on L-RB film as a reversible film.

Author Contributions: Data curation, A.F., Y.H., and T.K.; Supervision, T.M.; Writing—original draft, T.M.; Writing—review and editing, T.M. All authors have read and agreed to the published version of the manuscript.

Funding: This research received no external funding.

Acknowledgments: We would like to appreciate Toshiaki Hattori (Toyohashi University of Technology) for the donation of PHMB very much.

Conflicts of Interest: The authors declare no conflict of interest.

References

1. Abad-Villar, E.M.; Etter, S.F.; Thiel, M.A.; Hauser, P.C. Determination of chlorhexidine digluconate and polyhexamethylene biguanide in eye drops by capillary electrophoresis with contactless conductivity detection. *Anal. Chim. Acta* **2006**, *561*, 133–137. [CrossRef]
2. Masadome, T.; Hattori, T. Determination of polyhexamethylene biguanide hydrochloride. *Rev. Anal. Chem.* **2014**, *33*, 49–57. [CrossRef]
3. Kusters, M.; Beyer, S.; Kutscher, S.; Schlesinger, H.; Gerhartz, M. Rapid, simple and stability-indicating determination of polyhexamethylene biguanide in liquid and gel-like dosage forms by liquid chromatography with diode-array detection. *J. Pharm. Anal.* **2013**, *3*, 408–414. [CrossRef] [PubMed]
4. Lucas, D.; Gordon, E.A.; Stratmeyer, M.E. Analysis of polyhexamethylene biguanide in multipurpose contact lens solutions. *Talanta* **2009**, *80*, 1016–1019. [CrossRef] [PubMed]
5. Hattori, T.; Nakata, Y.; Kato, R. Determination of biguanide groups in polyhexamethylene biguanide hydrochloride by titrimetric methods. *Anal. Sci.* **2003**, *19*, 1525–1528. [CrossRef] [PubMed]
6. Hattori, T.; Tsurumi, N.; Kato, R.; Nakayama, M. Adsorptive voltammetry of 2-(5-Bromo-2-pyridyl) azo-5-[N-n-propyl-N-(3-sulfopropyl) amino]phenol on a carbon paste electrode in the presence of organic cations and polycation. *Anal. Sci.* **2006**, *22*, 1577–1580. [CrossRef] [PubMed]
7. Rowhani, T.; Lagalante, A.F. A colorimetric assay for the determination of polyhexamethylene biguanide in pool and spa water using nickel–nioxime. *Talanta* **2007**, *71*, 964–970. [CrossRef] [PubMed]
8. Ziolkowska, D.; Syrotynska, I.; Shyichuk, A. Quantitation of polyhexamethylene biguanide by photometric titration with Naphthol Blue Black dye. *Polimery* **2014**, *59*, 160–164. [CrossRef]
9. Masadome, T.; Yamagishi, Y.; Takano, M.; Hattori, T. Potentiometric titration of polyhexamethylene biguanide hydrochloride with potassium poly (vinyl sulfate) solution using a cationic surfactant-selective electrode. *Anal. Sci.* **2008**, *24*, 415–418. [CrossRef] [PubMed]
10. Masadome, T.; Miyanishi, T.; Watanabe, K.; Ueda, H.; Hattori, T. Determination of polyhexamethylene biguanide hydrochloride using photometric colloidal titration with crystal violet as a color indicator. *Anal. Sci.* **2011**, *27*, 817–821. [CrossRef] [PubMed]
11. Masadome, T.; Oguchi, S.; Kobayashi, T.; Hattori, T. Flow injection spectrophotometric Determination of polyhexamethylene biguanide hydrochloride in contact-lens detergents using Anionic Dyes. *Curr. Anal. Chem.* **2018**, *14*, 446–451. [CrossRef]
12. Uematsu, K.; Shimozaki, A.; Katano, H. Determination of polyhexamethylene biguanide utilizing a glucose oxidase enzymatic reaction. *Anal. Sci.* **2019**, *35*, 1021–1025. [CrossRef] [PubMed]
13. Masadome, T.; Asano, Y. Flow injection determination of cationic polyelectrolytes using a tetraphenylborate-selective electrode detector. *Bunseki Kagaku* **1999**, *48*, 515–518. [CrossRef]
14. Masadome, T.; Imato, T.; Itoh, S.; Asano, Y. Flow injection determination of anionic polyelectrolytes using an anionic surfactant-selective plasticized poly (vinyl chloride) membrane electrode detector. *Fresenius J. Anal. Chem* **1997**, *357*, 901–903. [CrossRef]
15. Masadome, T.; Akatsu, M. Optical sensor of anionic surfactants using solid-phase extraction with a lactone-form Rhodamine B membrane. *Anal. Sci.* **2008**, *24*, 809–812. [CrossRef] [PubMed]
16. Masadome, T.; Nakamura, Y.; Maruyama, K. Fluorescence optical sensing of anionic surfactants using a lactone-form Rhodamine B membrane. *Bunseki Kagaku* **2017**, *66*, 699–702. [CrossRef]
17. Sato, R.; Yamada, R.; Masadome, T. Development of an optode detector for determination of anionic surfactants by flow injection analysis. *Anal. Sci.* **2019**. [CrossRef] [PubMed]

Sample Availability: Not available.

© 2020 by the authors. Licensee MDPI, Basel, Switzerland. This article is an open access article distributed under the terms and conditions of the Creative Commons Attribution (CC BY) license (http://creativecommons.org/licenses/by/4.0/).

Article

A Fast and Validated High Throughput Bar Adsorptive Microextraction (HT-BAµE) Method for the Determination of Ketamine and Norketamine in Urine Samples

Samir M. Ahmad [1], Mariana N. Oliveira [1], Nuno R. Neng [1,2,*] and J.M.F. Nogueira [1,2,*]

[1] Centro de Química Estrutural, Faculdade de Ciências, Universidade de Lisboa, Campo Grande, 1749-016 Lisboa, Portugal; samir.marcos.ahmad@gmail.com (S.M.A.); mariananetoliveira@hotmail.com (M.N.O.)
[2] Departamento de Química e Bioquímica, Faculdade de Ciências, Universidade de Lisboa, Campo Grande, 1749-016 Lisboa, Portugal
* Correspondence: ndneng@fc.ul.pt (N.R.N.); nogueira@fc.ul.pt (J.M.F.N.)

Academic Editor: Paraskevas D. Tzanavaras
Received: 2 March 2020; Accepted: 17 March 2020; Published: 22 March 2020

Abstract: We developed, optimized and validated a fast analytical cycle using high throughput bar adsorptive microextraction and microliquid desorption (HT-BAµE-µLD) for the extraction and desorption of ketamine and norketamine in up to 100 urine samples simultaneously, resulting in an assay time of only 0.45 min/sample. The identification and quantification were carried out using large volume injection-gas chromatography-mass spectrometry operating in the selected ion monitoring mode (LVI-GC-MS(SIM)). Several parameters that could influencing HT-BAµE were assayed and optimized in order to maximize the recovery yields of ketamine and norketamine from aqueous media. These included sorbent selectivity, desorption solvent and time, as well as shaking rate, microextraction time, matrix pH, ionic strength and polarity. Under optimized experimental conditions, suitable sensitivity (1.0 µg L^{-1}), accuracy (85.5–112.1%), precision (≤15%) and recovery yields (84.9–105.0%) were achieved. Compared to existing methods, the herein described analytical cycle is much faster, environmentally friendly and cost-effective for the quantification of ketamine and norketamine in urine samples. To our knowledge, this is the first work that employs a high throughput based microextraction approach for the simultaneous extraction and subsequent desorption of ketamine and norketamine in up to 100 urine samples simultaneously.

Keywords: ketamine; norketamine; high throughput bar adsorptive microextraction; LVI-GC-MS(SIM); urine

1. Introduction

Ketamine (KET) was developed in 1962 during a search for a less problematic replacement for phencyclidine (PCP), an anesthetic that had gained notoriety for inducing hallucinations and psychosis. Due to its quick onset and short duration of action with only slight cardio-respiratory depression in comparison with other general anesthetics and the possibility of inhalation to maintain the anesthetic state, KET is a preferred drug for short-term surgical procedures in veterinary and human medicine, especially in children [1]. The main drawback of KET is its potential for causing vivid hallucinations, similar to those described for lysergic acid diethylamide (LSD) consumption [2]. As a result of this, it was initially abused by medical personnel and gradually became popular among young users at dance and rave parties [3]. In fact, the total quantity of KET seized worldwide increased from an annual average of 3 tons in the period 1998–2008 to 10 tons in the period 2009–2014 and 15 tons annually between 2015 and 2017 [4]. In humans, KET is metabolized in the liver by the microsomal cytochrome P450 system. CYP3A4 is the main enzyme responsible for KET N-demethylation to form norketamine (NKET). NKET is then hydroxylated, conjugated and excreted in the urine [5]. Studies of KET urinary excretion indicate that, over a 72-h period, little unchanged drug and NKET are present (2.3% and 1.6%, respectively). Most excreted compounds (80%) are conjugates of hydroxylated KET metabolites [6]. In urine collected from hospitalized children who had received KET as an anesthetic, it was detectable up to 2 days after drug administration (29–1410 µg L^{-1}) and NKET was detected for up to 14 days (up to 1559 µg L^{-1}) [7]. In a group of presumed recreational KET users, urine concentrations of KET and NKET were 6–7744 µg L^{-1} and 7–7986 µg L^{-1}, respectively [8].

Several analytical methods have been described in the literature for the determination of KET or NKET in urine samples, mostly using high performance liquid chromatography with ultraviolet/visible detection (HPLC-UV) [9], gas chromatography (GC) coupled to either flame ionization detector (FID) [10], mass spectrometry (MS) [3,7,11], *Tandem* mass spectrometry (MS/MS) [12] or even liquid chromatography (LC) coupled to MS [7,8] or MS/MS [13,14]. Sample preparation techniques used in combination with these chromatographic or hyphenated systems include conventional liquid-liquid extraction (LLE) [8] or solid-phase extraction (SPE) [7,14,15], but also miniaturized techniques such as solid-phase microextraction (SPME) [16], stir bar sorptive extraction [9], hollow-fiber liquid-phase microextraction [3,10] or microextraction by packed sorbent (MEPS) [12]. However, the preparation and manipulation of these techniques are tiresome, and the number of possible simultaneous microextractions are very limited, especially when routine work is involved. In order to overcome some of these issues we recently introduced an alternative approach, high-throughput bar adsorptive microextraction (HT-BAµE) [17]. This new technique and apparatus have shown to be user-friendly, cost-effective and presented remarkable effectiveness as a rapid tool for the simultaneous microextraction of up to 100 samples. In the present work, we propose the use of HT-BAµE for the extraction process and subsequent microliquid desorption of up to 100 samples in combination with large volume injection-gas chromatography-mass spectrometry operating in the selected-ion monitoring acquisition mode (HT-BAµE-µLD/LVI-GC-MS(SIM)) for the determination of KET and NKET in urine matrices. To our knowledge this is the first work that reports the use of a miniaturized high throughput methodology for the analysis of the target compounds in urine matrices.

2. Results and Discussion

2.1. Optimization Procedure

In order to maximize the extraction efficiencies for KET and NKET in aqueous media, different parameters affecting HT-BAµE-µLD/LVI-GC-MS(SIM) procedure have been investigated and optimized using a univariate approach, were the best values for each parameter were chosen for the next optimization assay, in accordance to previous similar works [17,18]. The initial conditions consisted in spiking 1.0 mL of ultrapure water (pH 5.5) with 100 µL of a working mixture containing KET and NKET resulting in a final concentration of 181.8 µg L^{-1}. Afterwards, the microextraction was performed for 90 min at 1000 rpm followed by µLD using 100 µL of MeOH containing 1.0 mg L^{-1} of IS under sonication (30 min, 42 +/− 2.5 kHz, 100 W). The parameters affecting the developed analytical approach were evaluated in a sequential order, starting from sorbent selectivity using 6 polymeric phases, which were chosen for their good performance for the extraction of polar to nonpolar compounds in aqueous media [17]. Next, the desorption solvent (MeOH, ACN and MeOH/ACN, 1/1, *v/v*; 100 µL) and time (from 5 to 60 min), as well as matrix pH (from 2.0 to 11.0), ionic strength (salt content from 0 to 20%, *w/v*) and polarity (organic modifier from 0 to 20%, *v/v*) were assayed. Finally, shaking speed (from 600 to 2200 rpm) and microextraction time (from 5 to 120 min) were evaluated.

Figure 1 depicts all the data from the optimization assays. The results clearly demonstrate that Strata-X presented higher recovery yields for KET and NKET than the other assayed sorbent coatings (Figure 1a). This result was expected since Strata-X promotes reverse-phase type mechanisms, such as π- and hydrophobic interactions, which normally favors the retention of non-polar compounds through its phenyl and polystyrene groups. Moreover, this sorbent coating also allows dipole-dipole interactions to occur through the n-specific action of the methoxyl and the carbonyl groups in the pyrrolidone group. Nevertheless, the hydrophobic character predominates and that is why the retention of non-polar compounds (PFE-06-MS (SD)). Taking into consideration the chemical characteristics (Table 1) of both KET and NKET analytes, the former appears to have mostly polar to ketone groups, resulting in semi-polar to non-polar characteristics (2.91 < log P < 3.35), which would favor its retention by Strata-X.

Table 1. Chemical structures, retention times (RT) and partition coefficients (log P) and acid dissociation constants (pKa) of IS, KET and NKET, and ions (of KET and NKET) obtained by LVI-GC–MS (SIM), under optimized instrumental conditions.

Analyte	Chemical Structure	log P [1]	pKa [1]	RT (min)	Ions (m/z) [2]
IS		-	-	5.42	83, 168, <u>169</u>
NKET		2.91	7.48	6.73	<u>166</u>, 168, 195
KET		3.35	7.45	6.97	<u>180</u>, 182, 209

[1] Calculator plugins were used for structure property prediction and calculation, Marvin 6.2.2, 2014, ChemAxon (http://www.chemaxon.com). [2] Quantification (underlined) and qualifier ions.

From the desorption optimization assays, it can be seen that no significant gain is achieved with either sorbent phase (Figure 1b). On the other hand, the optimum conditions for HLB were obtained by using 15 min of sonication time (Figure 1c) and 1 mL of MeOH:ACN (1:1, v/v) to desorb KET and NKET from the NVP-DVB sorbent using MeOH:ACN (MeCN):water (1:1, v/v) for 15 min.

The matrix pH, usually plays an important role in the microextraction process, where normally the non-ionized form of the target compounds seems to promote higher recovery yields from the aqueous media, since it favors the reverse-phase type interactions with the sorbent phases [17,18]. KET and NKET present weak basic characteristics (Table 1), being fully non-ionized at matrix pH > 9.5. For this reason, the recovery yields were maximized at the most alkaline pH assayed (pH 10)—Figure 1d.

By changing the solubility (Figure 1e) and the ionic strength (Figure 1f) of the aqueous matrix, the recovery greatly decreased for the former (especially with 20% MeOH, v/v) and that the recovery remained practically unchanged for the latter (with successive additions of NaCl). These results can be explained by the fact that increased matrix solubility normally favors the extraction of more non-polar compounds (log P > 3.5) and the increased ionic strength normally favors the extraction of very low polarity compounds (log P > 3.5) and the increased ionic strength normally favors the extraction of compounds with high polarity yield (log P < 2.0). KET and NKET present semi-polar to non-polar characteristics (2.91 < log P < 3.35), it would be expected that successive additions of NaCl or MeOH would probably hinder the microextraction process.

Finally, the effect of the shaking time (Figure 1g) and water extraction equilibrium time (Figure 1.h) for the extraction of KET and NKET in aqueous media water was also assayed. The data obtained through its phenyl and polystyrene groups. Moreover, this sorbent coating also allows dipole-dipole interactions, which normally favors the extraction of the more polar analytes through its pyrrolidone groups [19,20]. In general, these types of sorbents present higher selectivity for semi-polar to non-polar compounds (log P > 2.5). According to their chemical structures and polarity (Table 1), both KET and NKET present non-polar (phenyl) and polar (ketone) groups, resulting in semi-polar to non-polar characteristics (2.91 < log P < 3.35), which would favor its retention by Strata-X.

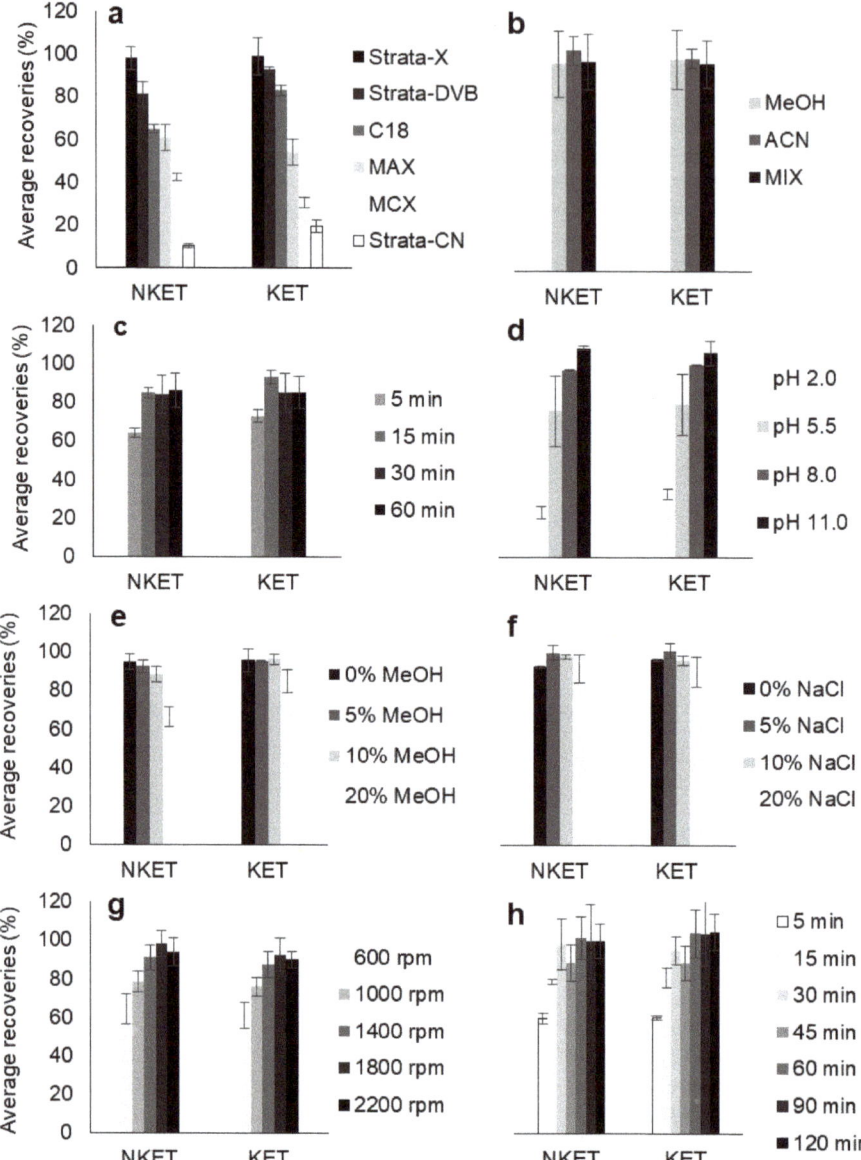

Figure 1. Effect of polymeric sorbent selectivity (**a**), microliquid desorption (μLD) solvent (**b**) and μLD time (**c**), matrix pH (**d**), polarity (**e**) and ionic strength (**f**), as well as shaking speed (**g**) and microextraction time (**h**) on the enrichment of ketamine (KET) and norketamine (NKET) from aqueous media, obtained by high-throughput bar adsorptive microextraction (HT-BAμE)-μLD/large volume injection-gas chromatography-mass spectrometry operating in the selected ion monitoring mode (LVI-GC-MS(SIM)). The error bars represent the standard deviation for the recovery levels of three replicates for each parameter evaluated. The microextraction devices were designed to be used only one time, once they are inexpensive and in order to avoid carryover effects [17].

Finally, the effect of the shaking speed (Figure 1g) and microextraction equilibrium time (Figure 1h) for the extraction of KET and NKET in aqueous media water was also assayed. The data obtained shows that 1800 rpm is the optimum value to microextract both target analytes by using only 30 min of, with no significant improvements using higher rates.

The method development resulted in the following optimized parameters: microextraction devices coated with Strata-X sorbent phase; extraction was performed for 30 min at 1800 rpm (pH 11.0); the µLD step was performed through cavitation (42 +/− 2.5 kHz, 100 W) for 15 min using 100 µL of the MeOH containing 1.0 mg L^{-1} of IS.

2.2. Validation Assays

The proposed methodology was validated following the parameters in accordance to Section 3.5 which included selectivity, linearity, sensitivity, accuracy, precision, as well as recovery yields and matrix effects. Table 2 shows most of the results for the validation results. Selectivity was assessed by verifying the absence of interfering peaks in the retention times of the target compounds using blank urine samples ($n = 10$). Each calibration plot showed good linearity ($r^2 \geq 0.999$; residuals $\leq 9.7\%$) over the range of 5.0 to 1000 µg L^{-1}. The linearity was also estimated (F_{calc}) using a lack-of-fit test (at confidence interval 95%) performed for both, which was always below the F_{tab}. As it can be observed the average recoveries yields and matrix effects using urine matrices at four spiking levels were between 84.9–105.0% (RSD $\leq 9.2\%$) and between −9.1–9.0% (RSD $\leq 14.1\%$), respectively. The accuracy values ranged from 87.2 to 110.0% (RSD $\leq 10.1\%$) and 85.5 to 112.1% (RSD $\leq 12.6\%$) for KET and NKET, respectively. These results show that the developed analytical approach is suitable for the analysis of KET and NKET in urine matrices.

2.3. Figures of Merit

In Table 3 we compare the LODS, linear range, accuracy, precision, recovery, sample volume and sample preparation time obtained by the proposed methodology and by other microextraction-based approaches. As it can be seen, the proposed work shows better sensitivity than most reported methodologies, even when using very sensitive instrumental systems such as GC-MS/MS [12]. The obtained LODs are only higher when compared to a methodology that uses larger amounts of sample volume [3]. The achieved accuracy, precision and recovery compares favorably with those depicted in Table 3, with the exception for a report using MEPS in combination with GC-MS/MS [12], although our proposed analytical approach uses much lower amounts of sample volume. Finally, as it can be seen, HT-BAµE-µLD/LVI-GC-MS(SIM), presents a much faster sample preparation time (0.45 min/sample) than the other microextraction-based methodologies using a high throughput configuration.

Figure 2 exemplifies total ion chromatograms from assays performed on spiked and unspiked urine sample, obtained by HT-BAµE-µLD/LVI-GC-MS(SIM), under optimized experimental conditions, where good selectivity and sensitivity are noticed, showing no endogenous interfering peaks at the retention times of the target compounds, including the IS.

Table 2. Intraday ($n = 6$) and interday ($n = 18$) accuracy (%) and precision (± relative standard deviation (RSD), %), recovery yields (% ± RSD, %) and matrix effects (% ± RSD, %) using four spiking levels, as well as limits of detection (LODs), lower limits of quantification (LLOQs), linear ranges and r^2, for KET and NKET in urine matrices, obtained by BAµE-µLD/LVI-GC-MS(SIM), under optimized experimental conditions.

Parameter	KET	NKET
LOD (µg L^{-1})	1.0	
LLOQ (µg L^{-1})	5.0	
Linear range (µg L^{-1})	5.0 to 1000.0	
Calibration plot ($n = 10$)	y = 0.0032x + 0.0066	y = 0.0032x + 0.029
r^2	0.9990	0.9970
Intra-day assays ($n = 6$)		
5.0 µg L^{-1}	87.2 ± 7.6	87.5 ± 11.9
50.0 µg L^{-1}	87.4 ± 6.6	98.8 ± 5.5
200.0 µg L^{-1}	87.9 ± 8.5	89.0 ± 6.8
1000.0 µg L^{-1}	94.8 ± 3.2	98.6 ± 4.5
Inter-day assays ($n = 18$)		
5.0 µg L^{-1}	110.0 ± 5.7	102.0 ± 12.6
50.0 µg L^{-1}	104.4 ± 10.1	112.1 ± 11.8
200.0 µg L^{-1}	94.7 ± 8.7	89.8 ± 12.3
1000.0 µg L^{-1}	102.9 ± 6.9	85.5 ± 6.1
Recovery yields ($n = 6$)		
5.0 µg L^{-1}	105.0 ± 9.2	103.1 ± 5.8
50.0 µg L^{-1}	97.8 ± 7.9	89.8 ± 4.7
200.0 µg L^{-1}	96.6 ± 7.2	88.1 ± 8.5
1000.0 µg L^{-1}	96.5 ± 4.0	84.9 ± 3.4
Matrix effect ($n = 6$)		
5.0 µg L^{-1}	−4.4 ± 6.1	8.4 ± 6.3
50.0 µg L^{-1}	4.9 ± 2.9	−4.6 ± 10.4
200.0 µg L^{-1}	9.0 ± 1.5	2.5 ± 14.1
1000.0 µg L^{-1}	−2.5 ± 6.2	−9.1 ± 5.6

Table 3. Comparison of the proposed method with other previously reported microextraction approaches for the determination of KET and NKET in urine samples.

Microextraction Technique	HF-LPME	MEPS	SBSE	SPME	HF-LPME	HT-BAµE
Instrumental system	GC-MS	GC-MS/MS	HPLC-UV	GC-MS	GC-FID	LVI-GC-MS
LODs(µg L^{-1})	0.1–0.25	5	2.3–9.1	100	8	1.0
Linear range(µg L^{-1})	0.5–50	10–250	30–3000	100–15000	3–350	5.0–1000.0
Accuracy (%)	88.3–108	91.4–105.6	n.a.	105.9–113.6	75.2–119.3	85.5–112.1
Precision (%)	≤10.1	≤9.2	≤8.9	≤14.8	≤8.9	≤12.6
Recovery (%)	85.2–101	72.5–100.7	90.8	n.a.	n.a.	84.9–105.0
Sample volume (mL)	2	0.25	3	1	3	0.5
Sample preparation time (min/sample)	60 [a]	7.42 [b]	40 [c]	21 [d]	20 [c]	45
Reference	[3]	[12]	[9]	[16]	[10]	This work

n.a. Information not available. [a] Multi-tube vortexer. Number of simultaneous microextractions not available. [b] 8 cycles of 500 µL, 1 cycle of 250 µL and 2 cycles of 100 µL at rates of 10.0 µL s^{-1}. [c] Magnetic stirrer. Number of simultaneous microextractions not available. [d] LEAP CombiPAL. Number of simultaneous microextractions not available.

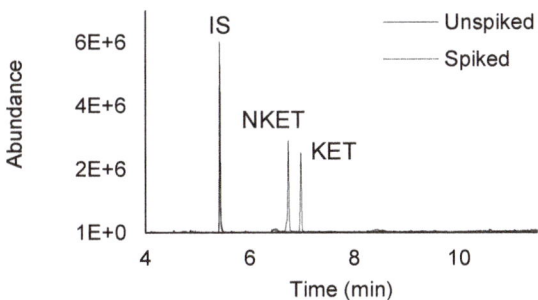

Figure 2. Total ion chromatogram of an assay from a spiked (125.0 μg L^{-1}) and unspiked urine sample, performed by HT-BAμE-μLD/LVI-GC-MS(SIM), under optimized experimental conditions.

Although the developed methodology was fully validated for the linear range of 5.0–1000.0 μg L^{-1}, KET and NKET was not detected (method) in the analyzed samples (for the) provided from 50 collected hospitalized children who had received KET rather anaesthetic, these detectable up to 2 days after drug administration (29.1–1.1 μg/L) had one spiked KET detected sample, it was a stable 1559 μg L^{-1}. The drug concentration relational KET and NKET was reported urine concentrations of KET and NKET were quantified in the range of a KET μg L^{-1}, and 3.79 for μg L urine [6], 7.2–87.3 and 5.3–5805 and NKET, 5.07-23031 and 5.87–8341 range of respectively in μg L^{-1}.7–7986 μg L^{-1} [8], 7.3–87.3 and 5.3–5805 μg L^{-1} [3], 5.07–23031 and 5.87–8341 μg L^{-1}, respectively [14].

3. Materials and Methods

The general sample preparation approach, chemicals, reagents and sorbent materials can already be found in the literature [17].

3.1. Chemicals, Sorbents and Samples

KET hydrochloride solution (1.0 mg mL^{-1} in MeOH), (±)-NKET hydrochloride solution (1.0 mg mL^{-1} in MeOH) and diphenylamine (internal standard, IS, 98.0%) were purchased from Sigma-Aldrich (Sigma-Aldrich, MI, USA). The di-sodium hydrogen phosphate anhydrous (Na$_2$HPO$_4$, 99.0%) from Panreac (Barcelona, Spain). Stock solutions of each standard were prepared at 100.0 mg L^{-1} by proper dilution with MeOH and stored at −20 °C in amber glass flasks and renewed every month. The standard mixtures used for method development and validation were prepared by appropriate dilution of the stock solutions in MeOH. The IS stock solution was prepared at 1.0 mg L^{-1}. Phosphate buffer (75.0 mmol L^{-1}, pH 11.0) was prepared by proper dilution of Na$_2$HPO$_4$ in ultra-pure water and by adding NaOH 1.0 mol L^{-1} until the desired solution pH was established (744 pH-meter, Metrohm, Herisau, Switzerland). All the stock solutions were stored light protected at 4 °C and renewed every week.

The authentic urine samples were provided by Joaquim Chaves Saúde clinic (Algés, Portugal). Upon arrival at the laboratory, the samples were frozen at −80 °C until use. For non-disclosure purposes, these samples were provided without any information from the donors. Blank urine samples used in all validation assays were obtained from our laboratory staff. It was specified that they could have consumed KET or any other related substances for at least a month before sampling. The study was approved by the Faculty Ethics Committee, authorization no. nr 4/2019.

3.2. LVI-GC-MS(SIM) Instrumentation

The GC-MS(SIM) instrumentation specifications can also be found in a previously published manuscript [21]. In this particular case, the injection conditions were as follows: vent time, 0.49 min; flow, 50 mL min^{-1}; pressure, 0 psi; purge flow, 12.9 mL min^{-1} at 2 min; the inlet temperature programmed from 80 °C (0.5 min) to 280 °C at a rate of 600 °C min^{-1}; 10 µL of injection volume at 100 µL min^{-1}. The oven temperature was programmed from 80 °C (held 1 min) to 200 °C at a rate of 50 °C min^{-1}, to 225 °C (held for 5 min) at a rate of 20 °C min^{-1}, to 250 °C at a rate of 20 °C min^{-1} and to 280 °C at a rate of 50 °C min^{-1} resulting in 11.5 min of total running time. The solvent delay was set at 4 min. For quantification purposes, calibration curves using the internal standard methodology were performed. For method optimization in ultra-pure water, relative peak areas obtained from each assay were compared with the relative peak areas of standard controls used for spiking. In Table 1 we present the retention times (RT) and ions (m/z) monitored for of KET, NKET and the IS obtained by LVI-GC-MS(SIM), under optimized instrumental conditions.

3.3. Pre-Treatment of Urine Samples

The urine samples were allowed to thaw and reach room temperature. The samples were vortexed for a few seconds, centrifuged for 10 min at 4500 rpm (Hermle Z 300, Germany) and the supernatants were collected. Afterwards, an acid hydrolysis was performed in order to obtain free KET or NKET from its corresponding conjugates, in accordance with the literature [15]. Therefore, 500 µL of the urine supernatants were pipetted into the microextraction vials already present in the HT-BAµE apparatus and 150 µL of HCOOH 10% (v/v) were added. Afterwards, the samples were heated to 40 °C for 1 h. After the samples were allowed to thaw and reach room temperature, 350 µL of phosphate buffer (75 mmol L^{-1}, pH 11.0) and 57.5 µL of NaOH solution (10 mol L^{-1}) were added in order to maintain pH 11.0. Finally, the vials were submitted to HT-BAµE-µLD analytical procedure.

The human urine samples were collected from voluntary donors with their informed consent.

3.4. HT-BAµE-µLD Methodology

After the pre-treatment step, the vials containing the samples were placed into the HT-BAµE apparatus, following a similar procedure already published but with a few alterations [17]. In the particular case, the BAµE devices were coated with NVP-DVB coating phase, the microextraction procedure was performed in an orbital shaker (Janke & Kunkel IKA-VIBRAX-VXR, Staufen, Germany) for 30 min at 1800 rpm, the microliquid desorption step was performed through cavitation (42 +/− 2.5 kHz, 100 W, Branson 3510, Carouge, Switzerland) for 15 min using 100 µL of the MeOH containing 1.0 mg L^{-1} of IS.

3.5. Validation of HT-BAµE-µLD/LVI-GC-MS(SIM) Methodology

Method validation was performed in accordance to similar reported analytical approaches [13,17]. The following parameters were studied: selectivity, linearity, sensitivity, accuracy and precision, as well as recovery and matrix effects. All validation assays were performed in triplicate, except when specified otherwise.

Sensitivity was assessed through the LOD and LLOQ. The former was calculated using a signal-to-noise ratio (S/N) of 3/1. The latter was determined as the lowest concentration values that was within acceptable accuracy and precision levels, i.e., the lowest calibration level.

The calibration plots (n = 10) were calculated using spiked blank urine samples ranging from 5.0 to 1000.0 µg L^{-1}. The linearity was estimated using lack-of-fit test, as well as by checking the respective determination coefficients (r^2) and residual plots.

Accuracy and precision were evaluated using quality control urine samples (QC) spiked with 5.0, 50.0, 200.0 and 1000.0 µg L^{-1}. Inter-day precision and accuracy were evaluated in three consecutive days. Precision was expressed as the RSDs (%) of the six assays for one day and eighteen assays for three consecutive days. Accuracy followed the same procedure but calculated as relative residuals (RRs) and was expressed as percent of the nominal concentration (%). The acceptance criterion for accuracy and precision was that RRs and RSDs should be ≤ 15.0%.

Matrix effect and average recovery assays were also determined concomitantly (n = 6). Average recovery yields were calculated as the ratio between the mean relative peak areas of the analytes obtained from QC before microextraction and samples spiked after microextraction using four concentration levels (5.0, 50.0, 200.0 and 1000.0 µg L^{-1}). Matrix effect was expressed as the ratio between the mean relative peak area obtained from QC spiked after microextraction and neat standard solutions at those same concentrations. Additionally, the RSDs of these two parameters were calculated to evaluate the variations that might arise from the matrix samples originating from different sources.

4. Conclusions

The methodology (HT-BAµE-µLD/LVI-GC-MS(SIM)) proposed in the present study, was fully optimized and validated to monitor KET and NKET in urine matrices. The proposed analytical cycle allowed to attain suitable analytical performance under optimized experimental conditions, including recovery, matrix effects, precision, accuracy, selectivity, sensibility and linear dynamic ranges. In addition to being user friendly, the proposed approach is environmentally friendly and cost-effective, once it takes into account the green analytical chemistry principles, i.e., uses only 100 µL of desorption solvent and 0.5 mL of urine sample per assay, does not require a derivatization step, and minimizes the overall time for the analytical procedure.

This analytical approach has the possibility of performing the microextractions and subsequent desorption of up to 100 samples in a single apparatus in just 45 min. This resulted in an average sample preparation time of 0.45 min/sample.

To our knowledge, this is the first work that employs a high throughput based microextraction approach for the simultaneous extraction and subsequent desorption of KET and NKET in up to 100 urine samples simultaneously.

Author Contributions: Conceptualization, S.M.A.; methodology, S.M.A.; software, S.M.A.; validation, S.M.A. and N.R.N.; formal analysis, S.M.A. and M.N.O.; investigation, S.M.A. and M.N.O.; resources, S.M.A. and M.N.O.; data curation, N.R.N.; writing—original draft preparation, S.M.A. and M.N.O.; writing—review and editing, N.R.N. and J.M.F.N.; visualization, N.R.N. and J.M.F.N.; supervision, J.M.F.N.; project administration, N.R.N.; funding acquisition, J.M.F.N. All authors have read and agreed to the published version of the manuscript.

Funding: This research was funded by Fundação para a Ciência e a Tecnologia (Portugal), grant numbers UIDB/00100/2020 and UIDP/00100/2020. S.M.A. PhD grant was funded by FCT (SFRH/BD/107892/2015). N.R.N. contract was funded by FCT through the contract established from DL 57/2016.

Acknowledgments: The authors thank Fundação para a Ciência e a Tecnologia (Portugal) for financial support through projects UIDB/00100/2020 and UIDP/00100/2020. The authors also thank FCT for the contract established from DL 57/2016 and the PhD grant (SFRH/BD/107892/2015). The authors also wish to thank Carlos Cardoso from Joaquim Chaves Saúde clinic (Algés, Portugal) for providing the urine samples.

Conflicts of Interest: The authors declare that they have no conflict of interest.

References

1. Dinis-Oliveira, R.J. Metabolism and metabolomics of ketamine: A toxicological approach. *Forensic Sci. Res.* **2017**, *2*, 2–10. [CrossRef] [PubMed]
2. Kuhar, M.J.; Liddle, H. *Drugs of Abuse*, 1st ed.; Marshall Cavendish Reference: New York, NY, USA, 2012; p. 173.
3. de Bairros, A.V.; Lanaro, R.; de Almeida, R.M.; Yonamine, M. Determination of ketamine, norketamine and dehydronorketamine in urine by hollow-fiber liquid-phase microextraction using an essential oil as supported liquid membrane. *Forensic Sci. Int.* **2014**, *243*, 47–54. [CrossRef] [PubMed]
4. World Drug Report 2019 (United Nations publication, Sales No. E.19.XI.8). Available online: https://wdr.unodc.org/wdr2019/ (accessed on 20 January 2020).
5. Clements, J.A.; Nimmo, W.S. Pharmacokinetics and analgesic effect of ketamine in man. *Br. J. Anaesth.* **1981**, *53*, 27–30. [CrossRef] [PubMed]
6. Karch, S. *Pathology of Drug Abuse*, 3rd ed.; CRC Press: Boca Raton, FL, USA, 2002; pp. 715–718.
7. Adamowicz, P.; Kala, M. Urinary excretion rates of ketamine and norketamine following therapeutic ketamine administration: Method and detection window considerations. *J. Anal. Toxicol.* **2005**, *29*, 376–382. [CrossRef] [PubMed]

8. Moore, K.A.; Sklerov, J.; Levine, B.; Jacobs, A.J. Urine concentrations of ketamine and norketamine following illegal consumption. *J. Anal. Toxicol.* **2001**, *25*, 583–588. [CrossRef] [PubMed]
9. Lan, L.; Hu, B.; Yu, C. PH-resistant titania hybrid organic-inorganic coating for stir bar sorptive extraction of drugs of abuse in urine samples followed by high performance liquid chromatography-ultraviolet visible detection. *J. Chromatogr. A* **2010**, *1217*, 7003–7009. [CrossRef] [PubMed]
10. Xiong, J.; Chen, J.; He, M.; Hu, B. Simultaneous quantification of amphetamines, caffeine and ketamine in urine by hollow fiber liquid phase microextraction combined with gas chromatography-flame ionization detector. *Talanta* **2010**, *82*, 969–975. [CrossRef] [PubMed]
11. Cheng, P.S.; Fu, C.Y.; Lee, C.H.; Liu, C.; Chien, C.S. GC-MS quantification of ketamine, norketamine, and dehydronorketamine in urine specimens and comparative study using ELISA as the preliminary test methodology. *J. Chromatogr. B Anal. Technol. Biomed. Life Sci.* **2007**, *852*, 443–449. [CrossRef] [PubMed]
12. Moreno, I.; Barroso, M.; Martinho, A.; Cruz, A.; Gallardo, E. Determination of ketamine and its major metabolite, norketamine, in urine and plasma samples using microextraction by packed sorbent and gas chromatography-tandem mass spectrometry. *J. Chromatogr. B Anal. Technol. Biomed. Life Sci.* **2015**, *1004*, 67–78. [CrossRef] [PubMed]
13. Hasan, M.; Hofstetter, R.; Fassauer, G.M.; Link, A.; Siegmund, W.; Oswald, S. Quantitative chiral and achiral determination of ketamine and its metabolites by LC–MS/MS in human serum, urine and fecal samples. *J. Pharm. Biomed. Anal.* **2017**, *139*, 87–97. [CrossRef] [PubMed]
14. Yang, C.A.; Liu, H.C.; Lin, D.L.; Liu, R.H.; Hsieh, Y.Z.; Wu, S.P. Simultaneous quantitation of methamphetamine, ketamine, opiates and their metabolites in urine by SPE and LC-MS-MS. *J. Anal. Toxicol.* **2017**, *41*, 679–687. [CrossRef] [PubMed]
15. Anilanmert, B.; Çavuş, F.; Narin, I.; Cengiz, S.; Sertler, Ş.; Özdemir, A.A.; Açikkol, M. Simultaneous analysis method for GHB, ketamine, norketamine, phenobarbital, thiopental, zolpidem, zopiclone and phenytoin in urine, using C18 poroshell column. *J. Chromatogr. B Anal. Technol. Biomed. Life Sci.* **2016**, *1022*, 230–241. [CrossRef] [PubMed]
16. Brown, S.D.; Rhodes, D.J.; Pritchard, B.J. A validated SPME-GC-MS method for simultaneous quantification of club drugs in human urine. *Forensic Sci. Int.* **2007**, *171*, 142–150. [CrossRef] [PubMed]
17. Ahmad, S.M.; Nogueira, J.M.F. High throughput bar adsorptive microextraction: A novel cost-effective tool for monitoring benzodiazepines in large number of biological samples. *Talanta* **2019**, *199*, 195–202. [CrossRef] [PubMed]
18. Abujaber, F.; Ahmad, S.M.; Neng, N.R.; Rodríguez Martín-Doimeadios, R.C.; Guzmán Bernardo, F.J.; Nogueira, J.M.F. Bar adsorptive microextraction coated with multi-walled carbon nanotube phases—Application for trace analysis of pharmaceuticals in environmental waters. *J. Chromatogr. A* **2019**, *1600*, 17–22. [CrossRef] [PubMed]
19. Ide, A.H.; Nogueira, J.M.F. New-generation bar adsorptive microextraction (BAμE) devices for a better eco-user-friendly analytical approach–Application for the determination of antidepressant pharmaceuticals in biological fluids. *J. Pharm. Biomed. Anal.* **2018**, *153*, 126–134. [CrossRef] [PubMed]
20. Ahmad, S.M.; Ide, A.H.; Neng, N.R.; Nogueira, J.M.F. Application of bar adsorptive microextraction to determine trace organic micro-pollutants in environmental water matrices. *Int. J. Environ. Anal. Chem.* **2017**, *97*, 484–498. [CrossRef]
21. Ahmad, S.M.; Gomes, M.I.; Ide, A.H.; Neng, N.R.; Nogueira, J.M.F. Monitoring traces of organochlorine pesticides in herbal matrices by bar adsorptive microextraction–Application to black tea and tobacco. *Int. J. Environ. Anal. Chem.* **2019**, 1–15. [CrossRef]

Sample Availability: Not available.

© 2020 by the authors. Licensee MDPI, Basel, Switzerland. This article is an open access article distributed under the terms and conditions of the Creative Commons Attribution (CC BY) license (http://creativecommons.org/licenses/by/4.0/).

Article

Partial Least Square Model (PLS) as a Tool to Predict the Diffusion of Steroids Across Artificial Membranes

Eleni Tsanaktsidou [1], Christina Karavasili [2], Constantinos K. Zacharis [1], Dimitrios G. Fatouros [2] and Catherine K. Markopoulou [1,*]

[1] Laboratory of Pharmaceutical Analysis, Department of Pharmacy, Aristotle University of Thessaloniki, 54124 Thessaloniki, Greece; etsanaktsi@pharm.auth.gr (E.T.); czacharis@pharm.auth.gr (C.K.Z.)
[2] Laboratory of Pharmaceutical Technology, Department of Pharmacy, Aristotle University of Thessaloniki, 54124 Thessaloniki, Greece; karavasc@pharm.auth.gr (C.K.); dfatouro@pharm.auth.gr (D.G.F.)
* Correspondence: amarkopo@pharm.auth.gr; Tel.: +30-231-099-7665

Received: 4 March 2020; Accepted: 16 March 2020; Published: 18 March 2020

Abstract: One of the most challenging goals in modern pharmaceutical research is to develop models that can predict drugs' behavior, particularly permeability in human tissues. Since the permeability is closely related to the molecular properties, numerous characteristics are necessary in order to develop a reliable predictive tool. The present study attempts to decode the permeability by correlating the apparent permeability coefficient (P_{app}) of 33 steroids with their properties (physicochemical and structural). The P_{app} of the molecules was determined by in vitro experiments and the results were plotted as Y variable on a Partial Least Squares (PLS) model, while 37 pharmacokinetic and structural properties were used as X descriptors. The developed model was subjected to internal validation and it tends to be robust with good predictive potential ($R^2Y = 0.902$, RMSEE = 0.00265379, $Q^2Y = 0.722$, RMSEP = 0.0077). Based on the results specific properties (logS, logP, logD, PSA and VD_{ss}) were proved to be more important than others in terms of drugs P_{app}. The models can be utilized to predict the permeability of a new candidate drug avoiding needless animal experiments, as well as time and material consuming experiments.

Keywords: steroids; Partial Least Squares regression; in vitro permeability; predictive model

1. Introduction

Steroids are an important category of active pharmaceutical ingredients (APIs). Their structure is characterized by a rigid steroid ring of cyclopentane-perhydro-phenanthrene or sterane [1]. Steroids are small lipophilic molecules and based on their genomic characteristics, they can enter the target cell by a passive diffusion mechanism (mainly by the transcellular route) across plasma membranes [2,3]. As they are derived from cholesterol, they are insoluble in water, and have many pharmacologic effects in almost every major system of the body including the endocrine, cardiovascular, musculoskeletal, nervous, and immune systems [4]. Due to their properties they can be administrated almost through every available administration route such as oral [5], buccal [6,7], transdermal [8], vaginal [9], otic [10], ocular [11], nasal [12], inhalation [13], intravenous [14].

A candidate drug should have appropriate physicochemical and pharmacological properties in order to successfully pass the pre-clinical and clinical trials. Such compound, must exhibit acceptable pharmacokinetic scheme in terms of absorption, distribution, metabolism, excretion and tolerable toxicity (ADMET). The simultaneous optimization of the above processes is one of the main challenges of current pharmacological research [15–17]. Unfortunately, these methods are laborious and extremely time-consuming, and they typically require 10–13 years [18,19].

Nowadays, there is a huge number of new candidate drugs that are designed and synthesized in the laboratory. In order to minimize the consumed cost and time, the pharmacokinetic behavior of the compound can be predicted using computational tools (e.g., cheminformatics) providing reliable pharmacokinetic models [15].

Quantitative structure activity/property relationship (QSAR/QSPR) studies correlate the physicochemical properties of a compound to biological activity [16]. Such studies have been extensively used for developing predictive models in which the chemical structures and biological properties are correlated. Alternately, such data could be obtained through in vitro, ex vivo and in vivo experiments [15].

Due to development of cheminformatics, there are plenty of QSAR modeling techniques, such as support vector machine (SVM), artificial neural networks (ANNs), multiple linear regression (MLR), principal component analysis (PCA) and partial least squares (PLS) regression [15]. The PLS method provides the possibility for linear correlation of numerous observations and multiple X variables with one or more Y variables [17].

Generally, PLS is a rapid and effective method for developing robust and reliable QSAR models. It has been widely used for the design of plenty of predictive patterns, such as for the placental-barrier permeability [18], blood–brain-barrier permeability from simulated chromatographic conditions [19], central nervous system (CNS) drug exposure [20], blood–brain barrier permeation of α-adrenergic and imidazoline receptor ligands using the parallel artificial membrane permeability assay (PAMPA) technique [21]. Additionally, PLS tool was used to discover potent Wee1 inhibitors [22], to evaluate 2-cyano-pyrimidine analogs as cathepsin-K inhibitors [23] and also to characterize the performance of dry powder inhalers [24].

The main scope of this research is to develop a new model that would be able to predict the permeability of a compound having the chemical structure of steroids. This approach is based on the correlation of its characteristics (physicochemical and structural properties) with the permeability of the molecule determined by in vitro experiments. In the present study the permeability of 33 steroids has been investigated using vertical Franz type diffusion cells including a synthetic cellulose membrane as model membrane [25]. Due to low water solubility of steroids, solubility enhancers (e.g., Polyethylene Glycol and Polysorbate 80) were used in order to achieve the desirable concentration for each compound. The obtained experimental results were treated using the Partial Least Squares (PLS) methodology. The developed models were validated and were found to be statistically significant with good predictive ability.

2. Results

2.1. Partial Least Squares (PLS) Methodology

2.1.1. Dataset Compilation

The present study involves the data processing of the derived experimental results using the PLS methodology. A Soft Independent Modeling of Class Analogies Simca-P (version 9; Umetrics, Uppsala, Sweden) [26,27] chemometric software was used to construct the classical PLS models.

The object of the research was to investigate the effect of several properties of steroids on their permeability at a hydrophilic cellulose membrane. The number of models developed in this process was five since the Y response variable was either calculated differently, or refers to four separate sampling times, after the first hour of the experiments. Therefore, P_{2h}, P_{4h}, P_{6h}, P_{8h} models denote the number of steroids permeating the artificial membrane at 2 h, 4 h, 6 h and 8 h, respectively (Y variable: permeability μg/cm^2), whereas *model* P_{app} expresses the (Y) variable calculated as the *apparent permeability* factor. In the present study, the theoretical explanation of steroids permeability was mainly based on model P_{app}, which is considered as the most important. Each of the five models contained 32 observations (analytes which belong to steroids) with 46 X variables and one Y variable. The large amount of X variables used was considered necessary, even though some of them were proved to be of minor interest. In order to implement the proposed models, it was rather important to carefully collect and record some of their most important properties and structural characteristics.

Each dataset consists of three parts. The first is the column containing the observations (33 analytes). The second is the main part of each dataset and it is populated by a few physicochemical and structural characteristics of the analytes. There are 37 descriptors (physicochemical properties), which were calculated using a series of different software or free online databases (Table 1).

Table 1. X Descriptors of Dataset.

Open Melting Point Dataset	EPA DSSTox	Data Warrior	ACD/Labs	Marvin	PubChem	pkCSM
Melting Point		1_cLogP	2_logP	3_logP	Topological PSA	MW
		1_cLogS	logD, pH 7.4	2_logS		4_LogP
		Hydrogen Bond Acceptors	Refractivity index	pKa (Strongest Acidic)		3_logS
		Hydrogen Bond Donors	Molar Refractivity	pKa (Strongest Basic)		Double bonds
		Aromatic Rings	PSA	HLB		Rotatable Bonds
		Carboxyl group	Polarizability	N$_o$ of Rings		Surface Area
		Carbonyl group	Molar Volume			Caco2 Permeability
		Hydroxyl group				Intestinal absorption
		Total Surface Area				log K$_p$
		Relative PSA				VD$_{ss}$
		PSA				log BB
		Shape Index				logPS
		Molecular Complexity				Total Clearance
		Molecular Flexibility				
		Drug-likeness				

In more details, the studied compounds were imported in the free cheminformatics program Data Warrior [28], in order to predict the clogP (calculated partition coefficient, log($C_{octanol}/C_{water}$)), the clogS (water solubility at 25 °C, log mol/L), the number of hydrogen bond acceptors and donors, the number of aromatic rings, carboxyl groups, carbonyl groups, hydroxyl groups, also the molecular complexity, the total surface area ($Å^2$), the relative polar surface area ($Å^2$), the polar surface area—PSA ($Å^2$), the shape index, the molecular flexibility and the drug-likeness. Descriptors related with the pharmacokinetic properties of the compounds were calculated by inserting the simplified molecular-input line-entry system (SMILES) of the drugs in the freely accessible web server pkCSM [29]. The pharmacokinetic properties employed were Caco2 permeability (log P_{app}), intestinal absorption (% absorption), skin permeability (log K_p), steady state of volume distribution (VD$_{ss}$, log L/kg), blood-brain barrier (BBB) permeability (logBB), CNS permeability (logPS), and total clearance (log mL/min/kg).

The melting point, °C of the compounds was obtained from Open Melting Point Dataset [30] and also from EPA DSSTox [31]. The topological polar surface area ($Å^2$) [32] was also predicted from PubChem data [33]. Moreover, Marvin, a free ChemAxon tool [34] was used in order to draw and characterize chemical structures of the compounds for the calculation of pK$_a$, logP, number of rings, distribution coefficient (logD) at pH 7.4 and their water solubility at 25 °C (logS, log mol/L). Details about the molar volume V$_m$ (cm^3), molar refractivity (cm^3), PSA ($Å^2$), polarizability (cm^3), molar volume (cm^3) were obtained via ACD/Labs [35]. All the above descriptors represented the X variables of the model developed and they are summarized in Table 1.

It is important to mention that some descriptors (e.g., logP) were calculated using more than one software program since there was a need to confirm their dominant role in the model. The structural features were found in the constitutional parameters and are outlined with nine descriptors used to decode the chemical structure of the analytes on the same basis. This was achieved by using integer numbers and zero to indicate the presence, the multiplicity or the absence of a structural characteristic.

The third part of the PLS dataset is a column with Y variable that corresponds to the calculated permeability of the drugs on Franz cells experiments. The Y variable is expressed as apparent permeability P_{app}, or permeability P_{2h}. P_{4h}. P_{6h}. P_{8h} at different sampling times (2h, 4h, 6h, 8h).

Variables' Importance in the Projection (VIP) column plots provide information about the importance of the parameters in the dataset. However, apart from the importance of a descriptor in a model, it is crucial to know whether its impact on the signal response is positive or negative. For this purpose, it was necessary to evaluate the loadings plots (w × c[1]/w × c[2]) of the models at the first two components.

2.1.2. Validation

Normalization of the observations (values of both X and Y variables) was achieved using mean centering and unit variance scaling. Validation of the PLS models was performed making use of three techniques, Cross-Validation (CV) the external and the internal validation [26,36].

First, the Cross Validation (CV) was achieved by dividing data into seven parts and each 1/7th of samples was excluded to build a model with the remaining 6/7th of samples. The Y values for the excluded data were then predicted by this new model and the procedure was repeated until all samples had been predicted once. If the original model is valid, then the plot of predicted Y versus actual measured Y values will be a straight line with the RMSEE (Root Mean Squares Error of Estimation) as low as possible (Figure 1) and calculated from Equation (1).

$$RMSEE = \sqrt{\frac{\sum (\hat{y}_i - y_i)^2}{N}} \qquad (1)$$

($\hat{y}_i$ represents the estimated P_{app} value for the ith object and y_i the reference P_{app} value)

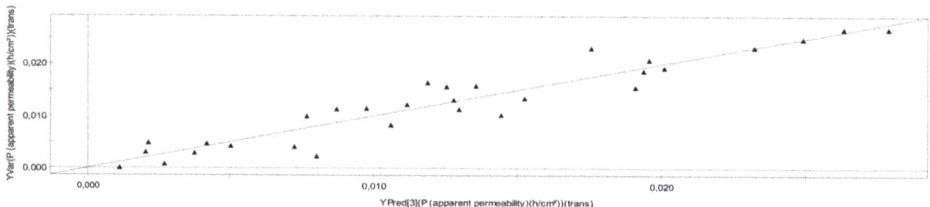

Figure 1. Observed versus estimated values of model P_{app} with *apparent permeability* values as (Y) variable, RMSEE = 0.00265379.

The prediction error sum of squares (PRESS) is a good measure of the predictive power of the model, providing information about the significance of the component (a component is considered significant when PRESS/Residual sum of squares < 1). Using the appropriate number of significant components, the total models were fit (Table 2) according to Haaland and Thomas criteria [37].

Table 2. Statistical Parameters in Partial Least Squares (PLS) regression models.

Models	R^2Y [1]	Q^2 (cum) [2]	Number of Components	Excluded Observations
P_{app}	0.902	0.722	3	3
P_{2h}	0.802	0.567	3	1
P_{4h}	0.847	0.656	3	2
P_{6h}	0.846	0.659	3	2
P_{8h}	0.872	0.605	3	3

[1] $R^2 = \sum_{i=1}^{N}(\hat{y}_i - y_i)^2 / \sum_{i=1}^{N}(y_i - \bar{y}_i)^2$ ($\bar{y}_i$ represents the means of the true P_{app} values in the predictor set).
[2] $Q^2 = \frac{1-PRESS}{SumSquares}$.

Verification of the reliability of the models was also achieved with the response permutation methodology (internal validation). During this process, the data for Y are not changed but they are randomly rearranged. Then the PLS model is applied again on the modified Y data and the R^2Y and Q^2Y values are recalculated. The above are compared with the initial values providing a first indication about the validity of the model. This process is repeated (20 permutations/model) and the results represent the statistical evaluation of the significance of the R^2Y and Q^2Y parameters in the initial model. In the diagram derived, the y-axis represents the R^2Y/Q^2Y values of all models and the x-axis represents the correlation coefficient between the modified and initial responses. In order to summarize the results of the method, regression analysis is applied on both R^2Y and Q^2Y and the regression lines are obtained. Verification of statistical significance of the original assessments is in accordance with the intercept limits regarding permutations (Figure 2) and they are set to $R^2Y < 0.3$–0.4 and $Q^2Y < 0.05$ [38].

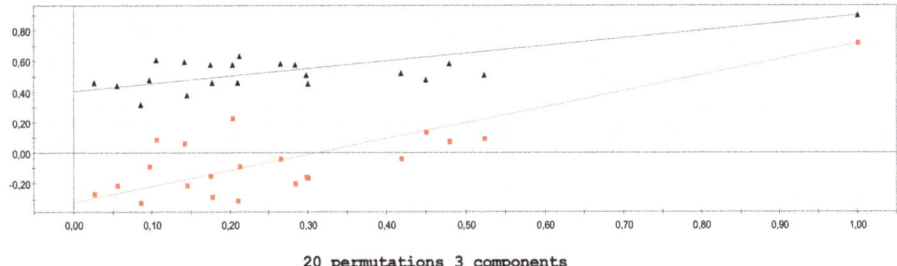

Figure 2. Internal validation test with 20 permutations, for model P_{app}. Intercepts: $R^2 = (0.0, 0.405)$ marked black, $Q^2 = (0.0, -0.32)$ marked red.

External validation was performed dividing data set of model P_{app} in two equal parts training and test set. Thereafter, the calculation of the training set and the prediction of the test were completed, and their roles were swapped. The quality of external prediction was assessed by the Q^2 ($Q^2_{train} = 75.4$, $Q^2_{test} = 71.5$) and the Root Mean Square Error of Prediction (RMSEP) from Equation (2) value, where RMSEP was equal to 0.00770361 for the training set and 0.00764925 for the test set, respectively.

$$RMSEP = \sqrt{\frac{\sum (obs - pred)^2}{N}} \quad (2)$$

The fact that the two estimates are similar means that these two subsets have similar information and can be combined in a total data set. External prediction may also aim the model to predict the Y values of new entities, in other words, entities excluded from the data set. Hence, the model is applicable either to interpret the behavior of a steroid based on its physicochemical properties or to predict the behavior of an unknown drug in the human body. PLS regression analysis is appropriate since it uses linear correlations and at the same time can predict with high reliability.

2.2. Interpretation of Steroids Permeability Through PLS

The permeability of a group of steroids across an artificial membrane was estimated using a hydrophilic cellulose membrane and the apparent permeability coefficient values were calculated. The mainly PLS model P_{app} was established using 32 compounds and a 47-descriptor analysis aimed at identification of the most critical molecular properties that influence permeability across the artificial membrane. According to the VIP plot of P_{app} model (Figure 3) logS, logP, logD (at pH 5.5 and 7.4), PSA (topological and relative) and VD_{ss} were found to be the most influential descriptors (VIP > 1) on the apparent permeability of the tested steroids through the cellulose membrane. All the other descriptors were found to have a similar and non-discriminating effect on the permeability of the tested compounds (VIP < 1).

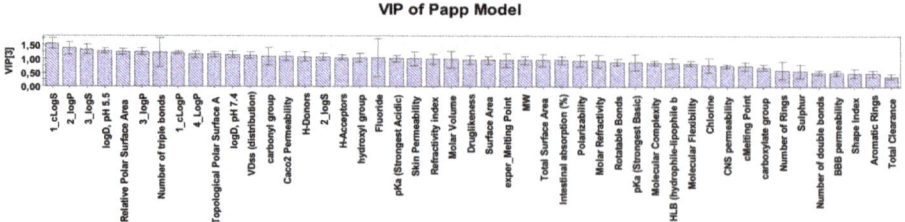

Figure 3. Variables' Importance in the Projection (VIP) plot for the *apparent permeability P* values of model P, at 95% confidence level.

Further information on the positive or negative effect of the X variables on the permeability is derived from scatter $w \times c[1]$ versus $w \times c[2]$ plot for P_{app} model in Figure 4.

Figure 4. A scatter $w \times c[1]$ versus $w \times c[2]$ plot for P_{app} model.

Drug dissolution is almost always a precondition for adequate permeability and absorption and, therefore, poor aqueous solubility is commonly associated with limited drug bioavailability [39]. It has been also exemplified that poor solubility may originate from high lipophilicity, resulting in poor permeability [40]. Compliant to this consensus, the findings of the current study support the positive contribution of logS (marked red in Table 3) and the respective negative effect of logP (marked blue in Table 3) on the P_{app} of the tested steroids.

Table 3. Models' VIP values.

Models' VIP Values							
P_{2h}		P_{4h}		P_{6h}		P_{8h}	
Var ID (Primary)	M2.VIP[3]	Var ID (Primary)	M3.VIP[3]	Var ID (Primary)	M3.VIP[3]	Var ID (Primary)	M3.VIP[3]
Total Clearance	1.62825	1_cLogS [1]	1.41597	1_cLogS	1.48033	1_cLogS	1.47640
Shape Index [2]	1.45142	Shape Index	1.31645	logD, pH 5.5 [3]	1.19879	No. of triple bonds	1.27167
Molar Volume	1.27798	Molar Volume	1.21737	3_logP	1.19154	Chlorine	1.22264
cMelting Point	1.25731	cMelting Point	1.18929	3_logS	1.18122	3_logS	1.18501
Refractivity index	1.21565	Molar Refractivity	1.15522	2_logS	1.17849	3_logP	1.17937
1_cLogS	1.21432	logD, pH 5.5	1.15097	Molar Volume	1.17612	Molar Volume	1.17567
Molar Refractivity	1.19812	3_logS	1.15097	Drug-likeness	1.16409	Drug-likeness	1.17289
Polarizability	1.17929	Polarizability	1.14560	cMelting Point	1.14712	Fluoride	1.16495
Total Surface Area	1.15973	2_logS	1.14311	Chlorine	1.13175	logD, pH 5.5	1.16166
No of triple bonds	1.14692	Chlorine	1.14308	Fluoride	1.12684	2_logS	1.15902
2_logS	1.12114	carboxylate group	1.13474	Molar Refractivity	1.12456	cMelting Point	1.13450
carboxylate group	1.11557	No of triple bonds	1.12947	Polarizability	1.11544	Molar Refractivity	1.11508
MW	1.09348	Fluoride	1.12462	1_cLogP	1.11174	Polarizability	1.10855
H-Donors	1.09223	Total Surface Area	1.12262	logD, pH 7.4	1.10117	Refractivity index	1.10584
Chlorine	1.08855	3_logP	1.12234	No of triple bonds	1.09925	Total Surface Area	1.08631
3_logS	1.07715	Refractivity index	1.10689	Total Surface Area	1.09019	1_cLogP	1.08316
logD, pH 5.5	1.05540	exper_Melting Point	1.07645	4_LogP	1.08878	hydroxyl group	1.07759
Rotatable Bonds	1.05147	MW	1.06777	Refractivity index	1.08517	4_LogP	1.06769
Surface Area	1.04603	H-Donors	1.05761	2_logP	1.08068	H-Donors	1.06767
exper_Melting Point	1.03503	logD, pH 7.4	1.05533	hydroxyl group	1.06590	logD, pH 7.4	1.06241
hydroxyl group	1.03056	1_cLogP	1.05320	H-Donors	1.06502	2_logP	1.0599
logD, pH 7.4	1.00956	hydroxyl group	1.04762	MW	1.04701	exper_Melting Point	1.05272
3_logP	0.97296	2_logP	1.04431	exper_Melting Point	1.03706	MW	1.04095
Caco2 Permeability	0.96943	Surface Area	1.04168	carboxylate group	1.03300	Surface Area	1.02086
4_LogP	0.96517	4_LogP	1.03678	Surface Area	1.02096	Shape Index	0.99368
1_cLogP	0.95059	Druglikeness	1.01070	Shape Index	0.99572	Relative PSA	0.98728
2_logP	0.93301	Rotatable Bonds	0.98079	Relative PSA	0.98111	carboxylate group	0.97549
H-Acceptors	0.88952	Caco2 Permeability	0.96573	Caco2 Permeability	0.96947	Rotatable Bonds	0.95868

[1] red indicates positive contribution, [2] green indicates size related descriptors, [3] blue indicates negative contribution.

PSA has been recently recognized as a useful predictor of permeability. It defines the polar part of a molecule and correlates with passive molecular transport through membranes. It has been previously observed that compounds with PSA < 60 Å are highly permeable, in contrast to those with PSA > 120 Å that are poorly permeable [41]. In that context, optimal permeability has been recognized when PSA is below 120 Å. Apart from prednisolone 21-sodium succinate (PSA = 141 Å), which has been classified as an outlier, PSA values for all steroids evaluated in the present study were below the cutoff value (PSA < 110 Å) suggested for the identification of poorly permeable compounds. Even though it's been recognized that lower PSA contributes to higher permeability [41], that trend was not confirmed here, mainly due to the absence of extreme variations in the PSA values and considering the relatively narrow range of PSA for the tested steroids.

Lipophilicity is considered one of the main factors with a positive effect on drug permeation across biological membranes. However, an inverse relationship between logP and permeability may be encountered upon increasing drug lipophilicity, due to a greater tendency for drug partitioning from the aqueous phase to the membrane [42]. It has been previously proposed that steroid permeation through a cellulose acetate membrane is a sequence of adsorption and desorption events with an intermittent membrane diffusion process, with the latter being dependent on the permeant's molecular size, its interaction with the membrane and the membrane's structural characteristics [43]. Such interaction might be favored with decrease in steroid polarity because, despite its hydrophilic nature, cellulose acetate remains more hydrophobic relative to the water [43]. Even though molecular size and polarity (with the latter being typically expressed as PSA or H-donors and acceptors) have been adversely associated with drug permeation [44,45], a positive correlation between steroids' polarity and permeability has been previously recognized. In particular, among three oestrogens of similar molecular size and distinctive polarities, an increase in permeability was observed with decreasing steroid-membrane interactions [46]. An inverse correlation between clogP and steroid permeability across Caco-2 cell monolayers was also recognized by Faasen et al., [47]. The results demonstrated a faster diffusion of the more hydrophilic steroids across the cell monolayers compared to the more hydrophobic ones. These findings coincide with the findings of the current study, concluding that steroids with lower logP gravitate towards a higher permeability.

Volume of distribution at steady state (VD_{ss}) is rendered as a solid indicator of drug distribution in the body reflecting its ability to permeate membranes and bind in tissues. Certain criteria have been defined to discriminate between drugs with high and low VD_{ss}. LogP has been shown to be a significant determinant of VD_{ss}, which along with the presence of Cl atoms and molecule compactness, have a positive contribution on this descriptor, while polarity and strong electrophiles have a negative contribution on VD_{ss} [48]. High VD_{ss} values (> 42 L), representative of more lipophilic drugs, indicate a high likeliness of drug distribution throughout body tissues, whereas low VD_{ss} values (< 3 L) associate with a predominant location in the systemic circulation [49]. According to the findings of the current study, a negative correlation between P_{app} and VD_{ss} was obtained, which aligns with the positive correlation between logP and VD_{ss} also observed in the present study.

Among the steroids evaluated, 4-chlorotestosterone demonstrated the lowest and prednisone and prednisolone the highest P_{app} value. As illustrated in Figure 5a, the presence of the Cl atom seems to be the most determinant descriptor affecting P_{app} of 4-chlorotestosterone. The chlorine atom as substituent in a molecule has been shown to increase its lipophilicity [19,50]. This positive contribution of Cl atoms to logP justifies the decrease in the apparent permeability of 4-chlorotestosterone, due to the negative correlation between logP and P_{app}, as already demonstrated in the present study. On the other hand, for both steroids showing the highest P_{app}, a combination of the same descriptors (logP and logS) was identified to be the most discriminative (Figure 5b,c). The steroids with the highest aqueous solubility and the lowest lipophilicity tend to diffuse faster across the cellulose membrane, compared to the most hydrophobic and less soluble steroids, which tend to a lower permeability.

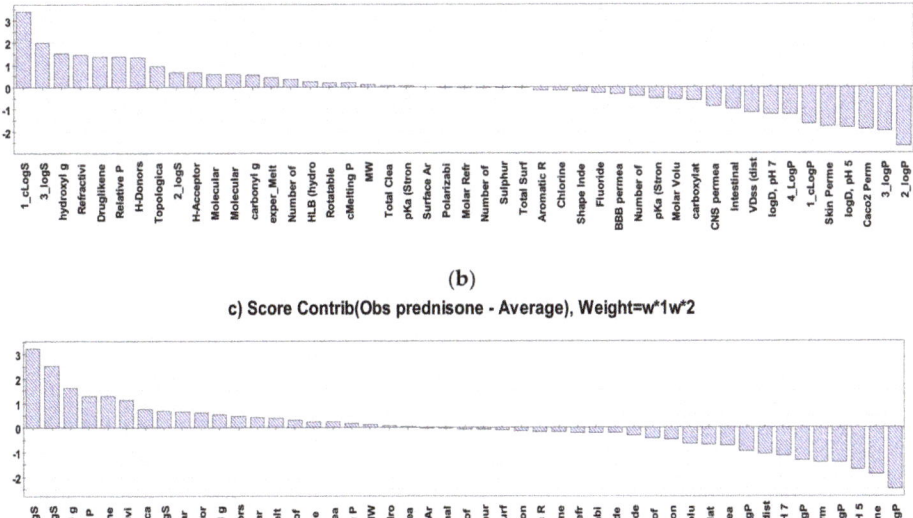

Figure 5. Contribution score plot of (**a**) 4-chlorotestosterone, (**b**) prednisolone and (**c**) prednisone, versus the remaining observations.

Additionally, androstanolone was considered as outlier during the first 4 h of the in vitro permeability study, showing significantly higher P_{app} values compared to the rest of the tested steroids. Based on its contribution plot at both 2 h and 4 h, it is evident that a combination of parameters related to the molecular size of androstanolone (number of double bonds, shape index, molar refractivity, polarizability, MW) are lower than the respective average values of the tested compounds, whereas pK$_a$ was found to be higher compared to the average pK$_a$ values of the tested steroids. Since all steroids remain unionized in the pH used in the current study, the contribution of pK$_a$ to P_{app} may be considered negligible. On the other hand, results signify the importance of molecular size on P_{app} with an inverse relationship existing between the two.

As already mentioned, drug diffusion across membranes consists of a series of events including drug transfer from the hydrophilic aqueous environment of the donor compartment, through the more hydrophobic (relative to the water) membrane to the hydrophilic aqueous environment of the receptor compartment. The ease of drug diffusion may be explained by elucidating the most significant parameters affecting drug permeability in a time-dependent manner. As seen in Table 3, the most critical descriptors affecting the amount of drug permeated at 2 h mainly relate to the molecular size of the permeants including shape index, MW and molar volume which is also directly related to the refractivity index and polarizability of the steroids tested [19], all the above are marked green in Table 3.

All these parameters contribute negatively to drug permeation, which could translate to hindering drug diffusion to the receptor phase and, instead, increasing their retention and affinity towards the membrane. This trend seems to change with time, with logS (red marked on Table 3) and logP (blue marked on Table 3) being the dominant descriptors affecting drug permeation thereafter.

The utility of in silico models in predicting drug permeability across biological membranes has been recognized as a time- and cost-efficient tool to facilitate drug discovery and development. The PLS model has been previously employed to identify the most critical molecular parameters affecting the permeability and retention of 17b-carboxamide steroids across an artificial membrane (parallel artificial membrane permeability assay (PAMPA)), as a means to predict their permeability across human skin [41]. In another study, Zhang et al., (2015) confirmed the good predictability of the PLS model, highlighting its potential utility as a high-throughput screening tool of placental drug permeability [18]. PLS and the genetic algorithm-PLS method have also been found appropriate in identifying the optimal subset of descriptors that have a significant contribution on drugs' permeability across Caco-2 cell monolayers [51], as well as on in vivo human drug intestinal permeability [52].

The present study is initially considered to be a reliable tool for the development of a theoretical background that will explain the permeability of steroids to biological membranes. In addition, the remarkable ability of PLS models to predict the behavior of drugs increases the usefulness of the proposed technique in designing new more effective steroids.

3. Materials and Methods

3.1. Reagents, Materials, Solutions

Acetonitrile (ACN) and water (HPLC grade) were purchased from VWR Chemicals (Radnor, USA), and Sigma-Aldrich (Darmstadt, Germany) respectively. For LC-MS analyses, water and ACN were both LC-MS gradient grade and provided by Sigma-Aldrich (Darmstadt, Germany).

Phosphate buffer saline (PBS) pH 7.4 was prepared by dissolving sodium chloride (8.0 g), sodium phosphate dibasic (1.44 g), and potassium phosphate monobasic (0.24 g), (Merck, Darmstadt, Germany) and potassium chloride (0.20 g) (Chem-Lab nv, Zedelgem, Belgium) in 1 L of distilled water. Polyethylene glycol (PEG 200) obtained by Sigma-Aldrich (Darmstadt, Germany) and polysorbate 80 (tween 80) provided by ManisChemicals (Athens, Greece).

The dialysis tubing cellulose membrane (flat width 43 mm) was obtained from Sigma-Aldrich (Darmstadt, Germany). The corticosteroids substances (Table 4) were United Stated Pharmacopeia—USP grade and were obtained from Sigma-Aldrich (Darmstadt, Germany).

Table 4. Structures of studied compounds.

	Compound	Double Bonds	C²	C³	C⁴	C⁵	C⁷	C⁹	C¹⁰	C¹¹	C¹³	C¹⁶	C¹⁷
17a-hydroxyprogesterone	COMP 1	4 = 5							CH_3		CH_3		$COCH_3$, OH
4-chlorotestosterone	COMP 2	4 = 5		=O	Cl				CH_3		CH_3		$OCOCH_3$
Androstanolone	COMP 3								CH_3		CH_3		OH
Betamethasone dipropionate	COMP 4	1 = 2, 4 = 5		=O				F	CH_3	OH	CH_3	CH_3	$COCH_2 OCOC_2H_5$, $OCOC_2H_5$
Betamethasone valerate	COMP 5	1 = 2, 4 = 5		=O				F	CH_3	OH	CH_3	CH_3	$COCH_2OH$, $OCOC_4H_9$
Budesonide	COMP 6	1 = 2, 4 = 5		=O					CH_3	OH		a	a, $COCCH_2OH$
Cortisone acetate	COMP 7	4 = 5		=O					CH_3	=O	CH_3		$COCH_2OCOCH_3$, OH
Dehydro-isoandrosterone	COMP 8	5 = 6		OH					CH_3		CH_3		=O
Deoxycorticosterone acetate	COMP 9	4 = 5		=O					CH_3		CH_3		$COCH_2OCOCH_3$
Dexamethasone	COMP 10	1 = 2, 4 = 5		=O				F	CH_3	OH	CH_3	CH_3	$CCOCH_2OH$, OH
D-norgestrel	COMP 11	4 = 5		=O							CH_2CH_3		C≡CH, OH
Estriol	COMP 12	1 = 2, 3 = 4, 5 = 10		OH								OH	OH
Estrone	COMP 13	2 = 3, 4 = 5, 10 = 1		OH							CH_3		=O
Ethinylestradiol	COMP 14	1 = 2, 3 = 4, 5 = 10		OH							CH_3		C≡CH, OH
Ethisterone	COMP 15	4 = 5		=O							CH_3	CH_3	C≡CH, OH
Fludrocortisone acetate	COMP 16	4 = 5		=O				F	CH_3	OH	CH_3		$COCH_2OCOCH_3$, OH
Formebolone	COMP 17	1 = 2, 4 = 5	CHO	=O					CH_3	OH	CH_3		CH_3, OH
Hydrocortisone	COMP 18	4 = 5		=O					CH_3	OH	CH_3		$COCH_2OH$, OH
Hydrocortisone acetate	COMP 19	4 = 5		=O					CH_3	OH	CH_3		$COCH_2OCOCH_3$, OH
Medroxyprogesterone acetate	COMP 20	4 = 5		=O		CH_3			CH_3		CH_3		$OCOCH_3$, $COCH_3$
Methandriol	COMP 21	5 = 6		OH									CH_3, OH
Methyl testosterone	COMP 22	4 = 5		=O						CH_3	CH_3		CH_3, OH
Norethisterone	COMP 23	4 = 5		=O							CH_3		C≡CH, OH
Prednisolone	COMP 24	1 = 2, 4 = 5		=O					CH_3	OH	CH_3		$COCH_2OH$, OH
Prednisolone 21-sodium succinate	COMP 25	1 = 2, 4 = 5		=O					CH_3	OH	CH_3		$COCH_2OCOCH_2CH_2OCOH$, OH
Prednisolone acetate	COMP 26	1 = 2, 4 = 5		=O					CH_3	OH	CH_3		$COCH_2OCOCH_3$, OH
Prednisone	COMP 27	1 = 2, 4 = 5		=O					CH_3	=O	CH_3		$COCH_2OH$, OH
Progesterone	COMP 28	4 = 5		=O					CH_3		CH_3		$COCH_3$
Spironolactone	COMP 29	4 = 5		=O			$SCOCH_3$		CH_3		CH_3		b
Testosterone	COMP 30	4 = 5		=O					CH_3		CH_3		OH
Testosterone acetate	COMP 31	4 = 5		=O					CH_3		CH_3		$OCOCH_3$
Testosterone propionate	COMP 32	4 = 5		=O					CH_3		CH_3		$OCOC_2H_5$
trans-Androsterone	COMP 33			OH					CH_3		CH_3		=O

sterane   a   b

3.2. Methods

3.2.1. Solubility Study

Solubility studies were carried out for the most lipophilic corticosteroid based on its logS value, obtained from Marvin (COMP 9, logS = −5.79, 25 °C). The study was conducted in PBS (pH 7.4) in the presence of polyethylene glycol 200 (PEG 200) and polysobrate 80 (Tween80) used as co-solvents at different ratios, owing the ability to enhance the water solubility of lipophilic drugs. In detail, an excess amount of the drug was added in the above-mentioned solvent mixtures and sonicated for 1 h at 30 °C. Then, the mixtures were kept under mild agitation for 48 h at room temperature to facilitate the dissolution. Any visible remaining drug particulates were removed by centrifugation at 2000× g for 20 min. The supernatants were quantified by HPLC analysis using the conditions described in

Table 5. Based on the results of the solubility study, PBS, 40 % (w/w) PEG 200 and 0.2 % (w/w) Tween 80 was used as the solvent mixture. The same procedure was followed for all compounds in this solvent mixture and a final concentration of 100 µg/mL was selected for the in vitro permeation studies.

Table 5. Chromatographic Conditions.

Compound	Detector	Flow (mL/min)	Retention Factor (k')	λ (nm)	Quantification Ion [2] (m/z)
COMP 1	DAD [1]	0.3	1.11	240	-
COMP 2	MS [3]	0.5	4.67	-	365 [(M+CH$_3$CN)+H]$^+$
COMP 3	MS	0.5	4.06	-	332 [(M+CH$_3$CN)+H]$^+$
COMP 4	DAD	0.3	2.18	240	-
COMP 5	DAD	0.5	1.45	240	-
COMP 6	DAD	0.5	1.93	240	-
COMP 7	DAD	0.4	3.05	240	-
COMP 8	DAD	0.4	2.55	230	-
COMP 9	DAD	0.4	2.43	240	-
COMP 10	DAD	0.4	2.46	240	-
COMP 11	DAD	0.4	3.34	240	-
COMP 12	DAD	0.5	1.49	205	-
COMP 13	DAD	0.5	3.98	205	-
COMP 14	DAD	0.4	0.92	240	-
COMP 15	DAD	0.3	2.56	205	-
COMP 16	DAD	0.4	3.12	240	-
COMP 17	DAD	0.4	2.58	220	-
COMP 18	DAD	0.3	0.83	240	-
COMP 19	DAD	0.5	0.96	240	-
COMP 20	DAD	0.5	0.74	240	-
COMP 21	MS	0.5	4.76	-	287 [(M-H$_2$O)+H]$^+$
COMP 22	DAD	0.5	0.87	240	-
COMP 23	DAD	0.4	2.48	240	-
COMP 24	DAD	0.4	2.18	240	-
COMP 25	DAD	0.4	3.11	244	-
COMP 26	DAD	0.4	2.43	240	-
COMP 27	DAD	0.4	2.45	240	-
COMP 28	DAD	0.5	0.87	240	-
COMP 29	DAD	0.5	4.04	238	-
COMP 30	DAD	0.5	0.67	240	-
COMP 31	DAD	0.5	1.12	240	-
COMP 32	DAD	0.5	1.66	240	-
COMP 33	MS	0.5	4.50	-	373 [(M+2CH$_3$CN)+H]$^+$

[1] DAD: diode array detector, [2] performed at single ion monitoring (SIM) mode, [3] MS: mass spectrometry.

3.2.2. In Vitro Permeation Studies

Cellulose membrane was properly treated and mounted in the Franz diffusion cells (diffusion area 4.9 cm^2, compartment volume 20 mL). The acceptor compartment was filled with PBS pH 7.4 and the donor compartment was filled with 1 mL of the formulation described above (100 µg/mL of the compounds). Permeation studies were conducted under constant stirring (90 rpm) at 37 °C. Samples of 0.5 mL were withdrawn from the acceptor compartment at predetermined time intervals (30 min, 1 h, 2 h, 4 h, 6 h, 8 h) and replaced with fresh and preheated PBS. Experiments were repeated in triplicates for each compound and blank experiments containing only the medium were also performed. The samples were analyzed by HPLC without any previous pretreatment.

Steady state flux (J_{ss}) was calculated from the slope of the linear section of the plot of the amount of permeated compound per unit area (µg/cm^2) against to time. The apparent permeability coefficient (P_{app}) was calculated using Equation (3), where C_d is the initial concentration of the drug in the donor compartment and J_{ss} is the steady state flux.

$$P_{app} = \frac{J_{ss}}{C_d} \quad (3)$$

3.2.3. HPLC Experimental Conditions/Method Validation

The drug content was quantified using either the HPLC-UV (High-performance Liquid Chromatography–Ultraviolet) or LC-MS (Liquid Chromatography–Mass Spectrometry) instrument. The HPLC-UV setup was equipped with two LC-20AD pumps, a SIL-20AC HT auto-sampler, a CTO-20AC column oven and an SPD-M20A diode array detector (Shimadzu). For LC-MS analysis, a Shimadzu LC-MS 2020 single-quadrupole mass spectrometer with an electrospray ion source (ESI) was utilized. A nitrogen gas generator N_2LCMS (Nitrogen Generator, Claind) was used throughout in this study. The temperature of the curved dessolvation line was set at 250 °C, the N_2 nebulizer gas flow was maintained at 1.5 L/min and the drying gas flow was set at 15 L/min, while the interface voltage was set at 4.5 kV in positive mode. The analytical column temperature was kept constant at 30 °C. The stationary phase was a C_{18} column (4.6 × 50 mm, 2.5 µm, Shimadzu). The sample injection volume was 5 µL in all cases. The mobile phase was a binary mixture of acetonitrile and water at appropriate ratio for each compound in order to avoid the prolonged analysis time. The HPLC-UV and LC-MS experimental conditions used for the analysis of each compound are described in Table 5.

Both the HPLC-UV and LC-MS methods were validated in-house according to ICH (International Conference Harmonization) guidelines [53]. The calibration curves for each compound was linear ($r^2 > 0.999$) in the range of LOQ-20 µg/mL (six calibration levels). Regression analysis, LOD (Limit of Detection) and LOQ (Limit of Quantification) values were tabulated in Table 6. Samples were analyzed in triplicate.

Table 6. Analytical figures of merit of HPLC-UV and LC-MS methods.

Compound	R^2	Intercept	Slope	LOD [1] (µg/mL)	LOQ [2] (µg/mL)
COMP 1	0.9997	1123	43133	0.01	0.04
COMP 2	0.9947	−5470	38034	0.27	0.90
COMP 3	0.9997	1252	16203	0.05	0.18
COMP 4	0.9997	−9501	28376	0.11	0.38
COMP 5	0.9998	−4479	20526	0.05	0.18
COMP 6	0.9998	552	16428	0.04	0.12
COMP 7	0.9996	−3723	28562	0.03	0.12
COMP 8	0.9997	−993	4454	0.32	1.07
COMP 9	0.9993	−4166	26523	0.02	0.06
COMP 10	0.9990	−11577	13079	0.18	0.60
COMP 11	0.9998	−12551	43421	0.01	0.03
COMP 12	1.0000	−1090	51548	0.07	0.24
COMP 13	0.9996	−3671	51988	0.06	0.19
COMP 14	0.9990	26763	79974	0.01	0.04
COMP 15	1.0000	706	30294	0.01	0.03
COMP 16	0.9999	7399	25115	0.04	0.14
COMP 17	0.9992	12588	45247	0.05	0.18
COMP 18	0.9997	−6597	62213	0.01	0.03
COMP 19	0.9985	26809	19890	0.03	0.10
COMP 20	1.0000	802	37460	0.01	0.05
COMP 21	0.9996	436	10500	0.05	0.16
COMP 22	0.9999	798	25856	0.02	0.05
COMP 23	0.9999	−162	38688	0.01	0.04
COMP 24	0.9999	1568	38840	0.03	0.10
COMP 25	1.0000	−353	13103	0.04	0.12
COMP 26	0.9997	−2111	22329	0.04	0.13
COMP 27	0.9997	−256	26564	0.04	0.14
COMP 28	0.9997	916	32027	0.01	0.04
COMP 29	1.0000	−1564	26756	0.02	0.06
COMP 30	1.0000	3599	113379	0.01	0.03
COMP 31	1.0000	6794	25375	0.01	0.04
COMP 32	1.0000	11227	54069	0.01	0.02
COMP 33	0.9971	−303	5860	0.03	0.10

[1] LOD: based on S/N = 3 criteria, [2] LOQ: based on S/N = 10 criteria.

4. Conclusions

An attempt to describe, experimentally and theoretically, the ability of a drug to permeate human tissues and be distributed in the body was carried out. For this purpose, five different PLS regression models were applied, using the permeability factor P_{app} as Y variable, for a series of steroids/drugs versus their physicochemical and structural properties (X variables). The determination of P_{app} factor was performed by in vitro drug permeability experiments across a cellulose membrane. According to the VIP values of the P_{app} model, the two factors with the stronger effect were logS and logP, which are dominant to the phenomenon with reverse influence. It is also remarkable that the permeability of steroids is dependent on the effect of numerous parameters and cannot be considered as a result of a specific factor (physicochemical property or structural feature). Finally, it is worth noting that one of steroids (4-chlorotestosterone) with chloro-substituted moiety did not penetrate the membrane at all, which makes it unique.

The PLS model seems to accurately describe this simulation and predict with reliability the behavior for an unknown drug. Based on such databases, researchers could use the information provided to predict whether a drug can be distributed in a tissue via passive transfer.

Author Contributions: Conceptualization, C.K.M.; Data curation, E.T.; Formal Analysis, E.T.; Investigation E.T. and C.K.M.; Methodology, C.K., D.G.F., C.K.Z. and C.K.M.; Supervision, C.K.M.; Validation, C.K.Z., C.K.M.; Writing and Original draft E.T., C.K., C.K.Z. and C.K.M.; Writing, Review and Editing, D.G.F. and C.K.M. All authors have read and agree to the published version of the manuscript.

Funding: This research received no external funding.

Conflicts of Interest: The authors declare no conflict of interest.

References

1. Alexander, K. Structure and Nomenclature of Steroids. In *Steroid Analysis*, 2nd ed.; July, G.D., Hugh, M., Eds.; Springer: Dordrecht, The Netherlands, 2010; pp. 1–25.
2. Giorgi, E.P.; Stein, W.D. The transport of steroids into animal cells in culture. *Endocrinology* **1981**, *108*, 688–697.
3. Oren, I.; Fleishman, S.J.; Kessel, A.; Ben-Tal, N. Free diffusion of steroid hormones across biomembranes: A simplex search with implicit solvent model calculations. *Biophys. J.* **2004**, *87*, 768–779.
4. McKay, C.J. Pharmacologic Effects of Corticosteroids. In *Holland-Frei Cancer Medicine*, 6th ed.; Kufe, D.W., Pollock, R.E., Weichselbaum, R.R., Eds.; BC Decker: Hamilton, ON, Canada, 2003.
5. Kuhnz, W. Pharmacokinetics of the contraceptive steroids levonorgestrel and gestodene after single and multiple oral administration to women. *Am. J. Obstet. Gynecol.* **1990**, *163*, 2120–2127.
6. Nair, M.; Chien, Y.W. Buccal delivery of progestational steroids: I. Characterization of barrier properties and effect of penetrant hydrophilicity. *Int. J. Pharm.* **1993**, *89*, 41–49.
7. Gass, M.; Rebar, R.W.; Cuffie-Jackson, C.; Cedars, M.I.; Lobo, R.A.; Shoupe, D.; Judd, H.L.; Buyalos, R.P.; Clisham, P.R. A short study in the treatment of hot flashes with buccal administration of 17-β estradiol. *Maturitas* **2004**, *49*, 140–147.
8. Badoud, F.; Boccard, J.; Schweizer, C.; Pralong, F.; Saugy, M.; Baume, N. Profiling of steroid metabolites after transdermal and oral administration of testosterone by ultra-high pressure liquid chromatography coupled to quadrupole time-of-flight mass spectrometry. *J. Steroid Biochem. Mol. Biol.* **2013**, *138*, 222–235.
9. Hassan, A.S.; Soliman, G.M.; El-Mahdy, M.M.; El-Gindy, G.E.-D.A. Solubilization and enhancement of ex vivo vaginal delivery of progesterone using solid dispersions, inclusion complexes and micellar solubilization. *Curr. Drug Deliv.* **2017**, *15*, 110–121.
10. Creber, N.J.; Eastwood, H.T.; Hampson, A.J.; Tan, J.; O'Leary, S.J. Adjuvant agents enhance round window membrane permeability to dexamethasone and modulate basal to apical cochlear gradients. *Eur. J. Pharm. Sci.* **2019**, *126*, 69–81.
11. Xu, X.; Sun, L.; Zhou, L.; Cheng, Y.; Cao, F. Functional chitosan oligosaccharide nanomicelles for topical ocular drug delivery of dexamethasone. *Carbohydr. Polym.* **2020**, *227*, 115356.

12. Guennoun, R.; Fréchou, M.; Gaignard, P.; Lière, P.; Slama, A.; Schumacher, M.; Denier, C.; Mattern, C. Intranasal administration of progesterone: A potential efficient route of delivery for cerebroprotection after acute brain injuries. *Neuropharmacology* **2019**, *145*, 283–291.
13. Demirca, B.P.; Cagan, H.; Kiykim, A.; Arig, U.; Arpa, M.; Tulunay, A.; Ozen, A.; Aydıner, E.K.; Baris, S.; Barlan, I. Nebulized fluticasone propionate, a viable alternative to systemic route in the management of childhood moderate asthma attack: A double-blind, double-dummy study. *Respir. Med.* **2015**, *109*, 1120–1125.
14. Zhuo, X.; Huang, X.; Yan, M.; Li, H.; Li, Y.; Rong, X.; Lin, J.; Cai, J.; Xie, F.; Xu, Y.; et al. Comparison between high-dose and low-dose intravenous methylprednisolone therapy in patients with brain necrosis after radiotherapy for nasopharyngeal carcinoma. *Radiother. Oncol.* **2019**, *137*, 16–23.
15. Ferreira, L.L.G.; Andricopulo, A.D. ADMET modeling approaches in drug discovery. *Drug Discov. Today* **2019**, *24*, 1157–1165.
16. Brown, E.W. © 1962 Nature Publishing Group. *Nat. Int. J. Sci.* **1962**, *196*, 1048–1050.
17. Wold, S.; Sjöström, M.; Eriksson, L. PLS-regression: A basic tool of chemometrics. *Chemom. Intell. Lab. Syst.* **2001**, *58*, 109–130.
18. Zhang, Y.H.; Xia, Z.N.; Yan, L.; Liu, S.S. Prediction of placental barrier permeability: A model based on partial least squares variable selection procedure. *Molecules* **2015**, *20*, 8270–8286.
19. Kouskoura, M.G.; Piteni, A.I.; Markopoulou, C.K. A new descriptor via bio-mimetic chromatography and modeling for the blood brain barrier (Part II). *J. Pharm. Biomed. Anal.* **2019**, *164*, 808–817.
20. Bergström, C.A.S.; Charman, S.A.; Nicolazzo, J.A. Computational prediction of CNS drug exposure based on a novel in vivo dataset. *Pharm. Res.* **2012**, *29*, 3131–3142.
21. Vucicevic, J.; Nikolic, K.; Dobričić, V.; Agbaba, D. Prediction of blood-brain barrier permeation of α-adrenergic and imidazoline receptor ligands using PAMPA technique and quantitative-structure permeability relationship analysis. *Eur. J. Pharm. Sci.* **2015**, *68*, 94–105.
22. Hu, Y.; Zhou, L.; Zhu, X.; Dai, D.; Bao, Y.; Qiu, Y. Pharmacophore modeling, multiple docking, and molecular dynamics studies on Wee1 kinase inhibitors. *J. Biomol. Struct. Dyn.* **2019**, *37*, 2703–2715.
23. Wang, J.; Peng, W.; Li, X.; Fan, W.; Wei, D.; Wu, B.; Fan, L.; Wu, C.; Li, L. Towards to potential 2-cyano-pyrimidines cathepsin-K inhibitors: An in silico design and screening research based on comprehensive application of quantitative structure-activity relationships, molecular docking and ADMET prediction. *J. Mol. Struct.* **2019**, *1195*, 914–928.
24. Elia, A.; Cocchi, M.; Cottini, C.; Riolo, D.; Cafiero, C.; Bosi, R.; Lutero, E. Multivariate data analysis to assess dry powder inhalers performance from powder properties. *Powder Technol.* **2016**, *301*, 830–838.
25. Ng, S.; Rouse, J.J.; Sanderson, F.D.; Eccleston, G.M. The Relevance of Polymeric Synthetic Membranes in Topical Formulation Assessment and Drug Diffusion Study. *Arch. Pharm. Res.* **2012**, *35*, 579–593.
26. Umetrics. *Simca-P 9.0—User Guide and Tutorial*; Umetrics: Malmo, Sweden, 2001.
27. Wu, Z.; Li, D.; Meng, J.; Wang, H. Introduction to SIMCA-P and Its Application. In *Handbook of Partial Least Squares: Concepts, Methods and Applications*; Vinzi, V.E., Chin, W.W., Henseler, J., Wang, H., Eds.; Springer: Berlin/Heidelberg, Germany, 2010; pp. 757–774.
28. Sander, T.; Freyss, J.; von Korff, M.; Rufener, C. DataWarrior: An open-source program for chemistry aware data visualization and analysis. *J. Chem. Inf. Model.* **2015**, *55*, 460–473.
29. Pires, D.E.V.; Blundell, T.L.; Ascher, D.B. pkCSM: Predicting small-molecule pharmacokinetic and toxicity properties using graph-based signatures. *J. Med. Chem.* **2015**, *58*, 4066–4072.
30. Bradley, J.-C.; Lang, A.; Williams, A. Jean-Claude Bradley Double Plus Good (Highly Curated and Validated) Melting Point Dataset. *Figshare* **2014**, *10*, m9.
31. Williams, A.J.; Grulke, C.; Edwards, J.; McEachran, A.D.; Mansouri, K.; Baker, N.; Patlewicz, G.; Shah, I.; Wambaugh, J.; Judson, R.S.; et al. The CompTox Chemistry Dashboard: A community data resource for environmental chemistry. *J. Cheminform.* **2017**, *9*, 61.
32. Ertl, P.; Rohde, B.; Selzer, P. Fast calculation of molecular polar surface area as a sum of fragment-based contributions and its application to the prediction of drug transport properties. *J. Med. Chem.* **2000**, *43*, 3714–3717.
33. Kim, S.; Chen, J.; Cheng, T.; Gindulyte, A.; He, J.; He, S.; Li, Q.; Shoemaker, B.A.; Thiessen, P.; Yu, B.; et al. PubChem 2019 update: Improved access to chemical data. *Nucleic Acids Res.* **2018**, *47*, D1102–D1109.
34. ChemAxon, Marvin. 2012. Available online: https://chemaxon.com/ (accessed on 9 September 2019).

35. ACD/Labs. Advanced Chemistry Development Inc. 2015. Available online: https://www.acdlabs.com/index.php (accessed on 9 September 2019).
36. Lapinsh, M.; Prusis, P.; Uhlén, S.; Wikberg, J.E.S. Improved approach for proteochemometrics modeling: Application to organic compound—Amine G protein-coupled receptor interactions. *Bioinformatics* **2005**, *21*, 4289–4296.
37. Haaland, D.M.; Thomas, E.V. Partial least-squares methods for spectral analyses. 1. Relation to other quantitative calibration methods and the extraction of qualitative information. *Anal. Chem.* **1988**, *60*, 1193–1202.
38. Lapins, M.; Eklund, M.; Spjuth, O.; Prusis, P.; Wikberg, J.E.S. Proteochemometric modeling of HIV protease susceptibility. *BMC Bioinform.* **2008**, *9*, 1–11.
39. Dahan, A.; Miller, J.M. The Solubility-permeability interplay and its implications in formulation design and development for poorly soluble drugs. *AAPS J.* **2012**, *14*, 244–251.
40. Tan, N.C.; Yu, P.; Kwon, Y.-U.; Kodadek, T. High-throughput evaluation of relative cell permeability between peptoids and peptides. *Bioorg. Med. Chem.* **2008**, *16*, 5853–5861.
41. Chi, C.T.; Lee, M.H.; Weng, C.F.; Leong, M.K. In silico prediction of PAMPA effective permeability using a two-QSAR approach. *Int. J. Mol. Sci.* **2019**, *20*, 3170.
42. Boyd, B.J.; Bergström, C.A.; Vinarov, Z.; Kuentz, M.; Brouwers, J.; Augustijns, P.; Brandl, M.; Bernkop-Schnürch, A.; Shrestha, N.; Préat, V.; et al. Successful oral delivery of poorly water-soluble drugs both depends on the intraluminal behavior of drugs and of appropriate advanced drug delivery systems. *Eur. J. Pharm. Sci.* **2019**, *137*, 104967.
43. Barry, B.W.; El Eini, D.I.D. Influence of non-ionic surfactants on permeation of hydrocortisone, dexamethasone, testosterone and progesterone across cellulose acetate membrane. *J. Pharm. Pharmacol.* **1976**, *28*, 219–227.
44. Palm, K.; Stenberg, P.; Luthman, K.; Artursson, P. Polar molecular surface properties predict the intestinal absorption of drugs in humans. *Pharm. Res.* **1997**, *14*, 568–571.
45. Veber, D.F.; Johnson, S.R.; Cheng, H.-Y.; Smith, B.R.; Ward, K.W.; Kopple, K.D. Molecular properties that influence the oral bioavailability of drug candidates. *J. Med. Chem.* **2002**, *45*, 2615–2623.
46. Barry, B.W.; Brace, A.R. Permeation of oestrone, oestradiol, oestriol and dexamethasone across cellulose acetate membrane. *J. Pharm. Pharmacol.* **1977**, *29*, 397–400.
47. Faassen, F.; Kelder, J.; Lenders, J.; Onderwater, R.; Vromans, H. Physicochemical Properties and Transport of Steroids across Caco-2 Cells. *Pharm. Res.* **2003**, *20*, 177–186.
48. Zhivkova, Z.; Mandova, T.; Doytchinova, I. Quantitative structure—Pharmacokinetics relationships analysis of basic drugs: Volume of distribution. *J. Pharm. Pharm. Sci.* **2015**, *18*, 515–527.
49. Smith, D.A.; Beaumont, K.; Maurer, T.S.; Di, L. Volume of distribution in drug design. *J. Med. Chem.* **2015**, *58*, 5691–5698.
50. Naumann, K. Influence of chlorine substituents on biological activity of chemicals: A review. *Pest. Manag. Sci.* **2000**, *56*, 3–21.
51. Yamashita, F.; Wanchana, S.; Hashida, M. Quantitative structure/property relationship analysis of Caco-2 permeability using a genetic algorithm-based partial least squares method. *J. Pharm. Sci.* **2002**, *91*, 2230–2239.
52. Winiwarter, S.; Ax, F.; Lennernäs, H.; Hallberg, A.; Pettersson, C.; Karlén, A. Hydrogen bonding descriptors in the prediction of human in vivo intestinal permeability. *J. Mol. Graph. Model.* **2003**, *21*, 273–287.
53. ICH. *ICH Topic Q2 (R1) Validation of Analytical Procedures: Text and Methodology*; ICH Secretariat: Geneva, Switzerland, 2005.

Sample Availability: Samples of all studied compounds are available from the authors.

© 2020 by the authors. Licensee MDPI, Basel, Switzerland. This article is an open access article distributed under the terms and conditions of the Creative Commons Attribution (CC BY) license (http://creativecommons.org/licenses/by/4.0/).

Communication

Chemometric Analysis of Low-field ^{1}H NMR Spectra for Unveiling Adulteration of Slimming Dietary Supplements by Pharmaceutical Compounds

Nao Wu, Stéphane Balayssac *, Saïda Danoun, Myriam Malet-Martino and Véronique Gilard *

Groupe de RMN Biomédicale, Laboratoire SPCMIB (UMR CNRS 5068), Université Paul Sabatier, Université de Toulouse, 118 route de Narbonne, 31062 Toulouse Cedex, France; nao.wu@chimie.ups-tlse.fr (N.W.); danoun@chimie.ups-tlse.fr (S.D.); martino@chimie.ups-tlse.fr (M.M.-M.)
* Correspondence: balayssac@chimie.ups-tlse.fr (S.B.); gilard@chimie.ups-tlse.fr (V.G.); Tel.: +33-5-61-55-82-81 (V.G.)

Academic Editor: Constantinos K. Zacharis
Received: 18 February 2020; Accepted: 3 March 2020; Published: 6 March 2020

Abstract: The recent introduction of compact or low-field (LF) NMR spectrometers that use permanent magnets, giving rise to proton (^{1}H) NMR frequencies between 40 and 80 MHz, have opened up new areas of application. The two main limitations of the technique are its insensitivity and poor spectral resolution. However, this study demonstrates that the chemometric treatment of LF ^{1}H NMR spectral data is suitable for unveiling medicines as adulterants of slimming dietary supplements (DS). To this aim, 66 DS were analyzed with LF ^{1}H NMR after quick and easy sample preparation. A first PLS-DA model built with the LF ^{1}H NMR spectra from forty DS belonging to two classes of weight-loss DS (non-adulterated, and sibutramine or phenolphthalein-adulterated) led to the classification of 13 newly purchased test samples as natural, adulterated or borderline. This classification was further refined when the model was made from the same 40 DS now considered as representing three classes of DS (non-adulterated, sibutramine-adulterated, and phenolphthalein-adulterated). The adulterant (sibutramine or phenolphthalein) was correctly predicted as confirmed by the examination of the ^{1}H NMR spectra. A limitation of the chemometric approach is discussed with the example of two atypical weight-loss DS containing fluoxetine or raspberry ketone.

Keywords: dietary supplement; adulteration; low-field NMR; multivariate analysis

1. Introduction

Nowadays, the adulteration of dietary supplements (DS) with approved or non-approved medicines poses a threat to the health of consumers [1]. The adulteration of slimming preparations with active pharmaceutical ingredients to increase their effects is a widely reported issue [2]. The two most common adulterants detected in weight-loss preparations are sibutramine and phenolphthalein, alone or in combination [3–5]. Sibutramine is an anorectic drug that was withdrawn from the market of many countries (European Union, USA, China, Australia, India …) since 2010 because of cardiovascular concerns. Phenolphthalein is used for its laxative properties even though it has been removed from over-the-counter products in the late 1990s due to a carcinogenic risk [6].

Various analytical methods have been proposed for the detection and/or quantification of undeclared drugs in slimming DS. The most frequently described technique is liquid chromatography with ultraviolet, diode-array or mass spectrometry detection [7–9]. Other methods including vibrational spectroscopy [10], gas chromatography [4] or ion exchange chromatography with conductivity detection have been proposed [11]. High-Field (HF) ^{1}H NMR spectroscopy has also been successfully applied for detecting and quantifying adulterants in slimming DS [5].

Low-field (LF) NMR is an emerging technique based on the use of a new generation of compact NMR [12–14]. A few applications of LF NMR in the pharmaceutical field have recently been described [15–19] and the feasibility of LF NMR for unveiling adulteration of DS has been demonstrated [15,19].

The aim of the present study is to deepen the evaluation of LF NMR to detect adulteration of slimming DS by coupling LF ^{1}H NMR data with a chemometric analysis, thus allowing classification of samples without expert interpretation of NMR spectra recorded on a low-cost benchtop spectrometer. We have thus analyzed adulterated and non-adulterated slimming DS, previously qualitatively and quantitatively characterized by HF ^{1}H NMR [5], with LF ^{1}H NMR in order to create statistical models in which the LF ^{1}H NMR data of new samples are injected. The interest and limitations of this approach are discussed.

2. Results and Discussion

2.1. LF ^{1}H NMR Analysis

The weight-loss DS used in this study, except the newly purchased test samples (T), were previously analyzed and fully characterized by HF ^{1}H NMR, i.e., the nature and amount of adulterants by unit (capsule, tablet, or sachet) were known [5]. The full list of DS is given in Table S1.

In the first step of the present study, all DS were analyzed in duplicate by LF ^{1}H NMR in deuterated methanol. The recording time of each spectrum was 15.5 min, and the profiles of typical samples are illustrated in Figure 1.

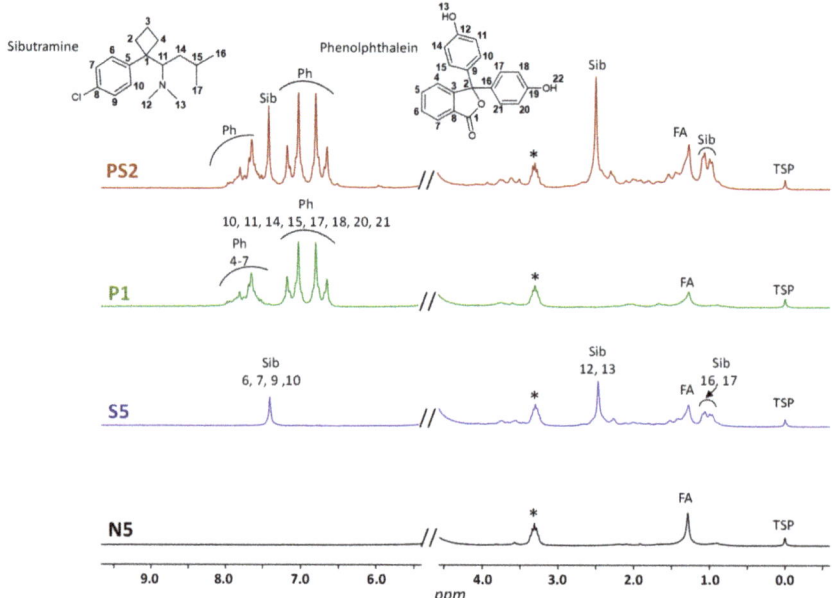

Figure 1. Typical LF ^{1}H NMR spectra of weight-loss dietary supplements recorded at 60 MHz (N, non-adulterated (natural) group; S, sibutramine-adulterated group; P, phenolphthalein-adulterated group; PS, both sibutramine and phenolphthalein-adulterated group). Ph: Phenolphthalein; Sib: Sibutramine; FA: Fatty acids; TSP: Internal reference; *: CD$_2$HOD.

Although LF ^{1}H NMR spectra are rather poorly resolved, the main characteristic signals of sibutramine and phenolphthalein, the two most common adulterants of slimming DS, are easily detected alone or in combination. As it can be seen in Figure 1, sibutramine is identified in samples S5 and PS2 by the signals of its aromatic protons at 7.41 ppm and of its methyl groups at 2.49 (CH_3 12 and 13) and 1.02 (CH_3 16 and 17) ppm. Likewise, aromatic protons of phenolphthalein give a characteristic pattern (6.5-8.0 ppm) that can be observed in DS P1 and PS2. Sample N5 is a DS without adulterant and, except for the reference and solvent signals, only the signal of some CH_2 protons of fatty acids from plant extracts is readily detected at 1.27 ppm. Minor signals corresponding to aromatic protons of natural polyphenols or other natural compounds are also detected in a few samples.

2.2. Chemometric Analysis

To start the chemometric analysis, a statistical model was built by performing a two-class comparison: DS without adulterant (natural: N, n = 19) were compared to DS containing either sibutramine (S, n = 12) or phenolphthalein (P, n = 9), samples (S) and (P) being considered together (n = 21) as "adulterated samples". After spectra processing (6–8 ppm region, see experimental part), bucketing and normalization of the data, the Partial Least Squares-Discriminant Analysis (PLS-DA) led to a predictive model with two principal PLS components and good validation criteria (Q^2 = 0.61, R^2Y = 0.76, CV-ANOVA = 2.3 × 10^{-18}). All Q^2 and R^2 values were lower in the permutation test than in the model, confirming its goodness. The classification of all samples was then obtained from the two-class model based on the predicted Y-values (YpredPS, which is the Y value predicted by the model based upon the X block variables (resonance intensities at given ppm)) indicating the probability that a sample belongs to one class of the model (adulterated or non-adulterated).

YpredPS values for the 66 DS analyzed in this study are reported in Figure 2. Samples (N), (S) and (P) (n = 40), whose content was previously known [5], were considered for the definition of a Y-value threshold between adulterated and non-adulterated DS. An YPredPS value close or superior to 1 would indicate that the sample is likely to belong to the adulterated class while an YPredPS value close to 0 would indicate that the sample is likely to be natural. Conventionally, a threshold of 0.65 was defined for samples belonging to a defined class and a 0.65–0.35 range for samples borderline to the defined class [20]. In our study, it appears that the adulterated samples (P) and (S) have YpredPS > 0.65, except P4 and P6 whose YpredPS are 0.30 and 0.32 respectively, whereas the highest YpredPS for (N) samples is 0.18. Based on the knowledge of the content of these two DS, we thus defined the lowest limit of the threshold at 0.30 (black dashed line in Figure 2). Samples (PS) previously shown as adulterated by both sibutramine and phenolphthalein were then injected into the model. Their YpredPS values being > 0.47, the upper limit of the threshold was set at 0.45 (red dashed line in Figure 2). So, DS with YpredPS values > 0.45 were considered as belonging to the adulterated class, those with YpredPS values < 0.30 to the non-adulterated class, and samples with YpredPS values between 0.30 and 0.45 were considered as borderline.

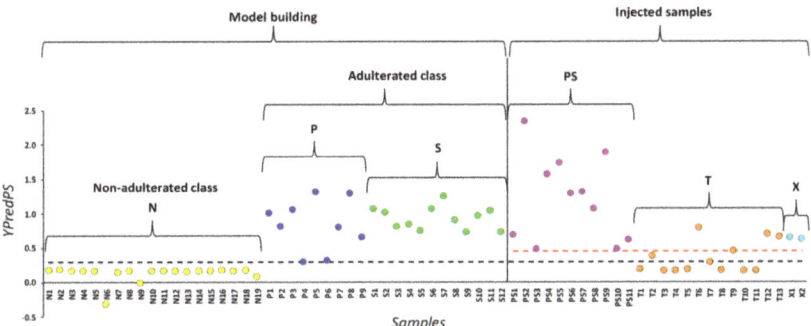

Figure 2. Predicted Y-values (YpredPS) obtained for the 66 DS analyzed based on the two-class PLS-DA model comparing natural samples (N) to adulterated samples (samples (P) and (S) considered together as a single class of adulterated samples). Samples above the red dashed line (YpredPS = 0.45) are defined as adulterated and those below the black dashed line (YpredPS = 0.30) as natural. PS, both sibutramine and phenolphthalein-adulterated group; T: test samples, i.e., newly purchased DS; X: atypical DS.

If we apply these criteria to the newly purchased DS (T, test samples), the classification shows that samples T6, T9, T12, and T13 are predicted adulterated, and samples T2 and T7 borderline, whereas other T samples are predicted natural with Y values ≤ 0.18 (Table 1). Two atypical DS, X1 ($YPredPS$ = 0.65) and X2 ($YPredPS$ = 0.63), which will be discussed later, are predicted adulterated by the two-class model. In conclusion, this preliminary rapid analysis with the two-class PLS-DA model can be considered as a first screening of adulterated slimming DS leading to a classification between natural, adulterated or possibly adulterated (borderline) samples.

Table 1. Classification list showing predicted Y-values (YPredPS) for test samples (T) based on the two-class PLS-DA model built with LF ^{1}H NMR data and completed by the visual observation of the projection of the samples on the three-class PLS-DA model shown in Figure 3A.

Identification	Predictive Y-value Classification from the Two-Class PLS-DA		Projection on the Three-Class PLS-DA Model Shown in Figure 3A	
	YPredPS	Classification	Class membership	Adulterant
T1	0.18	natural	N	-
T2	0.37	borderline	P	phenolphthalein
T3	0.16	natural	N	-
T4	0.17	natural	N	-
T5	0.18	natural	N	-
T6	0.79	adulterated	P	phenolphthalein
T7	0.30	borderline	P	phenolphthalein
T8	0.17	natural	N	-
T9	0.45	adulterated	S	sibutramine
T10	0.17	natural	N	-
T11	0.17	natural	N	-
T12	0.69	adulterated	S	sibutramine
T13	0.65	adulterated	S	sibutramine

To go further in the classification of the DS, a new PLS-DA analysis was carried out in which samples (N), (P) and (S) were considered as three distinct groups. A good predictive model was obtained with two principal PLS components (Q^2 = 0.66, R^2Y = 0.74), a p-value of the CV-ANOVA of 3.4×10^{-21}, and a permutation test successfully performed. The score plot of this three-class PLS-DA shows a clear discrimination between the three categories of DS (Figure 3A). Samples (P) (dark blue) and (S) (green) appear more spread out than samples (N) (yellow) because of the variable amount of adulterant in each sample ranging from 8 to 16 mg per unit for sibutramine in samples (S) and from 5 to 55 mg per unit for phenolphthalein in samples (P) [5].

(PS) samples (purple) projected in this three-class PLS-DA model are located closer to (P) than to (S) samples (Figure 3B). This observation is in agreement with higher amounts of phenolphthalein compared to sibutramine contained in most samples [5]. The score plot of the projection of test samples (T) in the model confirms the classification proposed in Table 1 but affords a more precise analysis (Figure 3C). Indeed, samples T1, T3–5, T8, T10, and T11 overlap with (N) samples and can thus be considered as natural. Samples T9, T12, and T13 contain the adulterant sibutramine whereas samples T2, T6, and T7 contain phenolphthalein. It can be noticed that none of the (T) samples belong to the (PS) class, i.e., contain a mixture of phenolphthalein and sibutramine. As the statistical analysis of the (T) samples was done blindly, i.e., without a thorough examination of their LF (and HF) ^{1}H NMR spectra, these findings were confirmed by the visual analysis of these spectra.

The fact that the two samples T2 and T7 were considered as borderline in the classification established from the predicted Y-values of the previous two-class model (Table 1) but are now better characterized by the three-class model (Figure 3C) can also be explained by the visual observation of their LF ^{1}H NMR spectra. Indeed, as reported in Figure 4, signals of phenolphthalein are detected in samples T2 and T7 but with a lower signal-to-noise ratio than in the P1 spectrum due to the low amount of adulterant in these DS. We mentioned above that signals corresponding to aromatic protons of natural polyphenols or other natural compounds were detected in a few (N) samples (as an illustration, the LF ^{1}H NMR spectrum of sample N6 is shown in Figure 4). Their chemical shifts and intensities close to those observed for phenolphthalein in T2 and T7 samples led to the classification of these DS as borderline in the first approach (Table 1).

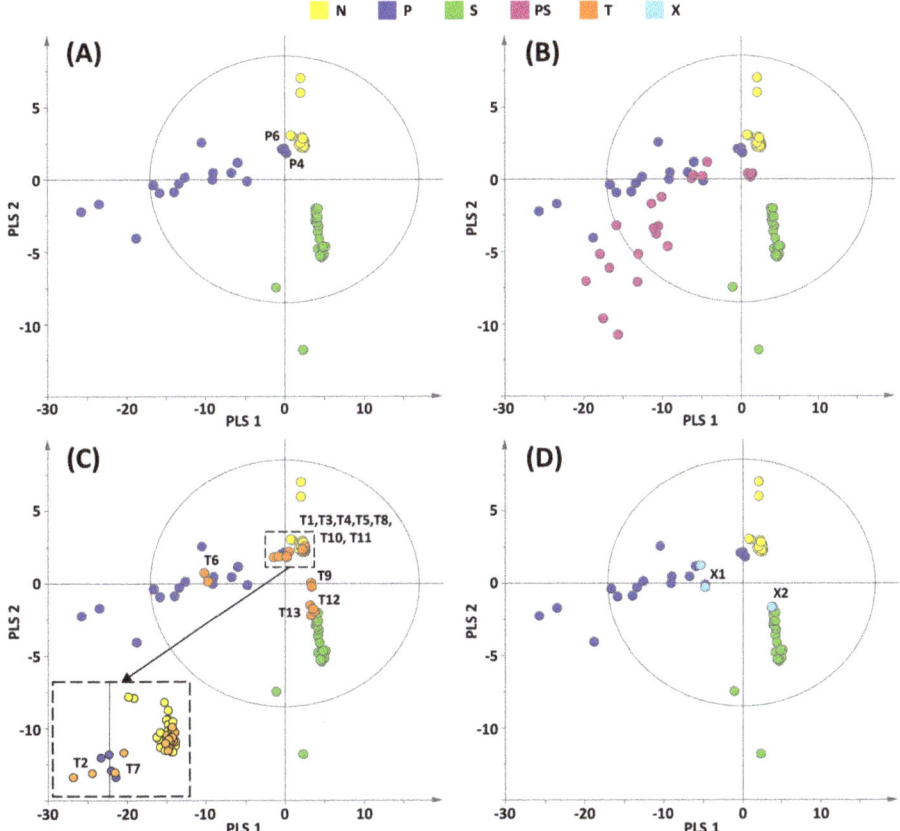

Figure 3. (**A**) Score plot of the PLS-DA three-class model built from LF ^{1}H NMR spectra of samples N (non-adulterated), S (adulterated with sibutramine), and P (adulterated with phenolphthalein). Score plots (**B**), (**C**) and (**D**) show the projection of samples PS (adulterated with both sibutramine and phenolphthalein), T (test samples) and X (atypical samples, see text) respectively on the built model (A).

A limitation of the present work is illustrated by the examples of the two DS X1 and X2. These samples appear as adulterated when considering their predicted Y-values (0.65 for X1 and 0.63 for X2) (Figure 2). Moreover, the projection of their LF ^{1}H NMR spectra in the PLS-DA three-class model indicates that the adulterants are phenolphthalein for X1 and sibutramine for X2 (Figure 3D). In fact, we demonstrated in a previous HF ^{1}H NMR study [5] that these two samples contain respectively raspberry ketone, a natural phenolic compound (probably intentionally added due to its high concentration in this particular DS), and fluoxetine, an antidepressant drug illegally added. As the model was built with only the LF ^{1}H NMR data of DS (N), (S) and (P), the chemometric analysis leads to the misclassification of X1 and X2. The reason can be found in their LF ^{1}H NMR spectra (Figure 4). Indeed, the main aromatic signal of fluoxetine has a chemical shift (7.37 ppm) close to that of sibutramine (7.41 ppm) and the large aromatic multiplet of raspberry ketone overlaps with the resonances of phenolphthalein. Although chemical shifts of these chemicals are slightly different from those of sibutramine or phenolphthalein, the multi-alignment procedure applied prior to the statistical treatment results in their misclassification. Nevertheless, even if the adulterant statistically identified in X2 is not the good one (i.e., sibutramine instead of fluoxetine), this DS remains unsafe for the consumer and the goal of the statistical screening for detecting dubious samples is thus achieved.

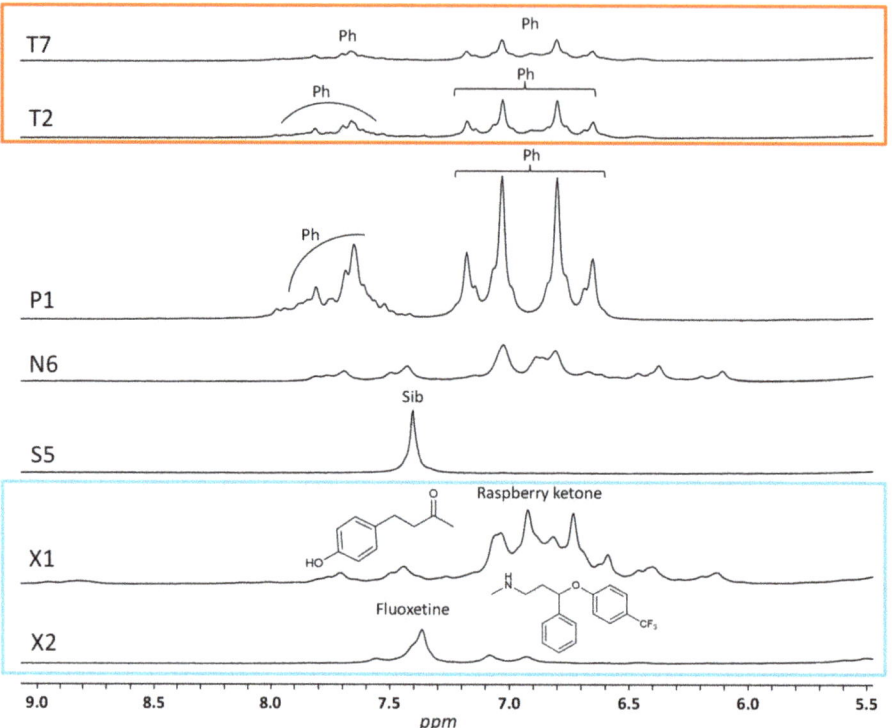

Figure 4. LF ^{1}H NMR spectra of some weight-loss dietary supplements recorded at 60 MHz. Ph: phenolphthalein; Sib: sibutramine.

In conclusion, the three-class PLS-DA works well as it enables a correct prediction of the nature of the adulterant sibutramine or phenolphthalein, the two banned drugs most commonly added to weight-loss DS to improve their effectiveness. The lowest limit of phenolphthalein concentration detected by the model is 3 mg per 100 mg of powder, which corresponds to ≈ 6 mg per unit if a mean capsule content weight of 200 mg is considered. The lowest limit of sibutramine concentration could not be reached because all the analyzed DS had YPredPS values > 0.7, very far from the 0.3 value that characterizes the limit between adulterated and non-adulterated DS (Figure 2). A source of classification error is nevertheless possible if an adulterant or a natural compound has a structure leading to ^{1}H NMR signals in the resonance frequency areas considered to build the model. For example, the characteristic signal of the methyl protons 16 and 17 of sibutramine at 1.02 ppm could not be used to create the model because it was often overlapped with the resonance of fatty acids.

This study shows that applying a chemometric treatment to LF ^{1}H NMR data is a means of widening the field of application of the technique, in particular for the analysis of complex mixtures. This approach has been successfully proposed in agri-food applications for the analysis and authentication of edible oils [21,22] and meat [23]. Very recently, a similar approach was used for the analysis of substandard and falsified medicines [19]. Our study expands the field to the adulteration of DS, a problem at the crossroads between agri-food and health products. In the case of slimming DS adulteration, the analytical process proposed can be useful for the first-line detection of samples liable to be adulterated without resorting to expert analysis of the ^{1}H NMR spectra. Sample preparation is simple and fast, and LF ^{1}H NMR acquisition is easy, quite push-button and does not require specific NMR knowledge. The perspective of this study would be to automate the whole process in order to propose a turnkey method that could be implemented in quality control labs.

3. Materials and Methods

3.1. Samples

Different groups of weight-loss DS were ordered on the Internet and analyzed with LF ^{1}H NMR: (N) without adulteration, (S) adulteration with sibutramine, (P) adulteration with phenolphthalein, (PS) adulteration with both phenolphthalein and sibutramine, (T) test samples, and (X) two atypical samples (Table S1). The 40 DS used for building the statistical models ((N) ($n = 19$), (S) ($n = 12$), and (P) ($n = 9$)) as well as (PS) samples ($n = 11$), and the two DS (X) were previously qualitatively and quantitatively characterized in our lab with HF ^{1}H NMR [5]. For testing the statistical models, 13 new DS (T1–T13) were bought on the Internet in November 2019 and were analyzed by LF and HF ^{1}H NMR upon receipt.

3.2. Sample Preparation for LF ^{1}H NMR Analysis

Around 100 mg of powdered samples were mixed with 1 mL of deuterated methanol under vortex agitation for 15 s and then sonicated for 5 min. The suspension was then centrifuged (5 min, 3000 rpm) and the supernatant (700 µL) analyzed. Thirty microliters of sodium 2,2,3,3-tetradeutero-3-(trimethylsilyl) propanoate (TSP, 40 mM) as the internal chemical shift reference was added before the NMR analysis. Each DS was prepared in duplicate.

3.3. LF ^{1}H NMR Analysis

Qualitative LF ^{1}H NMR spectra were acquired on a Pulsar™ benchtop NMR spectrometer (Oxford Instruments, Abingdon, UK) operating at a frequency of 59.7 MHz for ^{1}H. The temperature inside the spectrometer was 310 K. The acquisition was performed by using the SpinFlow 1.2.0.1 software (Oxford Instruments) and the processing was done with MNova 11.0 (Mestrelab Research, Santiago de Compostela, Spain). Free induction decays (FIDs) were recorded with a flip angle of 90° (12.5 µs), a spectral width of 5000 Hz (83.75 ppm), and 8 K complex points (acquisition time of 1.64 s). The relaxation delay was set at 2 s, and 256 transients were recorded leading to a total acquisition time of 15.5 min. For data processing, the FIDs were apodized with an exponential filter (line broadening (LB) of 0.3 Hz), and a Whittaker smoother was applied for automatic baseline correction. The number of points was increased to 16 K in Fourier transformed spectra. The signal of TSP set at 0 ppm was used as an internal reference for chemical shift (δ) measurement.

3.4. Chemometrics

First, the data matrix with all LF ^{1}H NMR spectra (132 spectra) of ((N), (S), (P), (PS), (T), and (X)) groups was generated in the Chemometrics module included in the MNova software with a spectral resolution of 0.01 ppm/point. Data were transferred to the Matlab® software (R2018a, The Mathworks Inc., Natick, MA, USA) for the alignment procedure using the Icoshift algorithm [24] with the following input arguments: PS9 as reference spectrum (target vector), data matrix with all spectra, a file for local alignment with three specific intervals 8.385–7.495, 7.495–7.275, and 7.275–6.055 ppm, an optional 'f' command for a fast search of the best alignment for each interval and no co-shift preprocessing step. Then, the bucketing procedure was performed with an optimized bucketing algorithm [25] and a fixed bin width of 0.01 ppm. Data were normalized by dividing their areas by that of the internal standard TSP signal and by the weight of powder in capsules, tablets, sachet, coffee or tea bags. Multivariate statistical analyses were done with the SIMCA-P+ 13.0 software (Umetrics, Umea, Sweden). PLS-DA with UV-scaling analyses were performed with two (80 spectra corresponding to 19 samples (N) and 21 adulterated samples (P) and (S)) or three qualitative variables (19 (N), 9 (P) and 12 (S)). Then (PS), (T) and (X) LF ^{1}H NMR data (52 spectra) were projected into the active model and predicted score plots were built. Predicted Y-values (YPredPS) were provided by the classification list included in the predict module of the SIMCA-P+ software.

Supplementary Materials: The following are available online at http://www.mdpi.com/1420-3049/25/5/1193/s1. Table S1. Information on slimming dietary supplements analyzed in this study.

Author Contributions: Conceptualization, V.G., S.B.; methodology, S.B., S.D.; validation, S.B., V.G.; formal analysis, S.B., N.W.; investigation, N.W., S.D.; writing—Original draft preparation, V.G., S.B.; writing—Review and editing, M.M.-M.; visualization, N.W., V.G., S.B.; supervision, V.G., S.B.; funding acquisition, M.M-M., V.G. All authors have read and agreed to the published version of the manuscript.

Funding: This research was funded by the French National Agency for the Safety of Medicines and Health Products (Agence Nationale de Sécurité du Médicament et des produits de santé: ANSM), grant AAP-2012-082, (convention ANSM/UPS n°2012S071).

Acknowledgments: The authors thank the Chinese Scholarship Council (CSC) for Ph.D. funding.

Conflicts of Interest: The authors declare no conflict of interest.

References

1. Ekar, T.; Kreft, S. Common risks of adulterated and mislabeled herbal preparations. *Food Chem. Toxicol.* **2019**, *123*, 288–297. [CrossRef] [PubMed]
2. Skalicka-Wozniak, K.; Georgiev, M.I.; Orhan, I.E. Adulteration of herbal sexual enhancers and slimmers: The wish for better sexual well-being and perfect body can be risky. *Food Chem. Toxicol.* **2017**, *108*, 355–364. [CrossRef] [PubMed]
3. Rooney, J.S.; McDowell, A.; Strachan, C.J.; Gordon, K.C. Evaluation of vibrational spectroscopic methods to identify and quantify multiple adulterants in herbal medicines. *Talanta* **2015**, *138*, 77–85. [CrossRef] [PubMed]
4. Xia, Z.; Cai, W.; Shao, X. Rapid discrimination of slimming capsules based on illegal additives by electronic nose and flash gas chromatography. *J. Sep. Sci.* **2015**, *38*, 621–625. [CrossRef] [PubMed]
5. Hachem, R.; Assemat, G.; Martins, N.; Balayssac, S.; Gilard, V.; Martino, R.; Malet-Martino, M. Proton NMR for detection, identification and quantification of adulterants in 160 herbal food supplements marketed for weight loss. *J. Pharm. Biomed. Anal.* **2016**, *124*, 34–47. [CrossRef] [PubMed]
6. Dunnick, J.K.; Hailey, J.R. Phenolphthalein exposure causes multiple carcinogenic effects in experimental model systems. *Cancer Res.* **1996**, *56*, 4922–4926.
7. Rebiere, H.; Guinot, P.; Civade, C.; Bonnet, P.A.; Nicolas, A. Detection of hazardous weight-loss substances in adulterated slimming formulations using ultra-high-pressure liquid chromatography with diode-array detection. *Food Addit. Contam. A* **2012**, *29*, 161–171. [CrossRef]
8. Kim, H.J.; Lee, J.H.; Park, H.J.; Cho, S.H.; Cho, S.; Kim, W.S. Monitoring of 29 weight-loss compounds in foods and dietary supplements by LC-MS/MS. *Food Addit. Contam. A* **2014**, *31*, 777–783. [CrossRef]
9. Zeng, Y.; Xu, Y.; Kee, C.L.; Low, M.Y.; Ge, X. Analysis of 40 weight loss compounds adulterated in health supplements by liquid chromatography quadrupole linear ion trap mass spectrometry. *Drug Test. Anal.* **2016**, *8*, 351–356. [CrossRef]
10. Deconinck, E.; Cauwenbergh, T.; Bothy, J.L.; Custers, D.; Courselle, P.; De Beer, J.O. Detection of sibutramine in adulterated dietary supplements using attenuated total reflectance-infrared spectroscopy. *J. Pharm. Biomed. Anal.* **2014**, *100*, 279–283. [CrossRef]
11. Dal Molin, T.R.; da Silveira, G.D.; Leal, G.C.; Muller, L.S.; Muratt, D.T.; de Carvalho, L.M.; Viana, C. A new approach to ion exchange chromatography with conductivity detection for adulterants investigation in dietary supplements. *Biomed. Chromatogr.* **2019**, *33*, e4669. [CrossRef] [PubMed]
12. Blumich, B.; Singh, K. Desktop NMR and its applications from Materials Science to Organic Chemistry. *Angew. Chem. Int. Ed. Engl.* **2018**, *57*, 6996–7010. [CrossRef] [PubMed]
13. Blumich, B. Low-field and benchtop NMR. *J. Magn. Reson.* **2019**, *306*, 27–35. [CrossRef] [PubMed]

14. Van Beek, T.A. Low-field benchtop NMR spectroscopy: Status and prospects in natural product analysis. *Phytochem. Anal.* **2020**, 1–14. [CrossRef]
15. Pages, G.; Gerdova, A.; Williamson, D.; Gilard, V.; Martino, R.; Malet-Martino, M. Evaluation of a benchtop cryogen-free low-field ^{1}H NMR spectrometer for the analysis of sexual enhancement and weight loss dietary supplements adulterated with pharmaceutical substances. *Anal. Chem.* **2014**, *86*, 11897–11904. [CrossRef]
16. Zivkovic, A.; Bandolik, J.J.; Skerhut, A.J.; Coesfeld, C.; Prascevic, M.; Zivkovic, L.; Stark, H. Quantitative analysis of multicomponent mixtures of over-the-counter pain killer drugs by Low-Field NMR spectroscopy. *J. Chem. Educ.* **2017**, *94*, 121–125. [CrossRef]
17. Assemat, G.; Gouilleux, B.; Bouillaud, D.; Farjon, J.; Gilard, V.; Giraudeau, P.; Malet-Martino, M. Diffusion-ordered spectroscopy on a benchtop spectrometer for drug analysis. *J. Pharm. Biomed. Anal.* **2018**, *160*, 268–275. [CrossRef]
18. Assemat, G.; Balayssac, S.; Gerdova, A.; Gilard, V.; Caillet, C.; Williamson, D.; Malet-Martino, M. Benchtop low-field ^{1}H Nuclear Magnetic Resonance for detecting falsified medicines. *Talanta* **2019**, *196*, 163–173. [CrossRef]
19. Keizers, P.H.J.; Bakker, F.; Ferreira, J.; Wackers, P.F.K.; van Kollenburg, D.; van der Aa, E.; van Beers, A. Benchtop NMR spectroscopy in the analysis of substandard and falsified medicines as well as illegal drugs. *J. Pharm. Biomed. Anal.* **2020**, *178*, 112939. [CrossRef]
20. Sandasi, M.; Vermaak, I.; Chen, W.; Viljoen, A. The application of vibrational spectroscopy techniques in the qualitative assessment of material traded as Ginseng. *Molecules* **2016**, *21*, 472. [CrossRef]
21. Parker, T.; Limer, E.; Watson, A.D.; Defernez, M.; Williamson, D.; Kemsley, E.K. 60 MHz ^{1}H NMR spectroscopy for the analysis of edible oils. *Trends Analyt. Chem.* **2014**, *57*, 147–158. [CrossRef] [PubMed]
22. Zhu, W.; Wang, X.; Chen, L. Rapid detection of peanut oil adulteration using low-field nuclear magnetic resonance and chemometrics. *Food Chem.* **2017**, *216*, 268–274. [CrossRef] [PubMed]
23. Jakes, W.; Gerdova, A.; Defernez, M.; Watson, A.D.; McCallum, C.; Limer, E.; Colquhoun, I.J.; Williamson, D.C.; Kemsley, E.K. Authentication of beef versus horse meat using 60 MHz ^{1}H NMR spectroscopy. *Food Chem.* **2015**, *175*, 1–9. [CrossRef] [PubMed]
24. Savorani, F.; Tomasi, G.; Engelsen, S.B. Alignment of 1D NMR data using the ICOSHIFT tool: A tutorial. In *Magnetic Resonance In Food Science*; Van Duynhoven, J., Belton, P.S., Webb, G.A., van As, H., Eds.; Royal Society of Chemistry: Cambridge, UK, 2013; pp. 14–24.
25. Sousa, S.A.A.; Magalhaes, A.; Ferreira, M.M.C. Optimized bucketing for NMR spectra: Three case studies. *Chemom. Intell. Lab. Syst.* **2013**, *122*, 93–102. [CrossRef]

Sample Availability: Samples of dietary supplements are available from the authors.

© 2020 by the authors. Licensee MDPI, Basel, Switzerland. This article is an open access article distributed under the terms and conditions of the Creative Commons Attribution (CC BY) license (http://creativecommons.org/licenses/by/4.0/).

MDPI
St. Alban-Anlage 66
4052 Basel
Switzerland
Tel. +41 61 683 77 34
Fax +41 61 302 89 18
www.mdpi.com

Molecules Editorial Office
E-mail: molecules@mdpi.com
www.mdpi.com/journal/molecules